A Practical Guide
for the
BEGINNING FARMER

by
Herbert Jacobs

DOVER PUBLICATIONS, INC.
NEW YORK

OTHER BOOKS BY HERBERT JACOBS

Try and Stump Me! Yearbook
We Chose the Country
The Community Newspaper
Middleton (Wis.) Centennial Book (Ed.)
Practical Publicity
Frank Lloyd Wright—America's Greatest Architect

Published in Canada by General Publishing Com-
pany, Ltd., 30 Lesmill Road, Don Mills, Toronto,
Ontario.
Published in the United Kingdom by Constable
and Company, Ltd., 10 Orange Street, London
WC2H 7EG.

This Dover edition, first published in 1978, is a
republication of the work originally published by
Harper & Brothers, New York, in 1951.
The Acknowledgments of the original edition
are here deleted, and a new Foreword and Bibliography have been prepared by the author especially
for the Dover edition.

International Standard Book Number: 0-486-23675-7
Library of Congress Catalog Card Number: 78-52593

Manufactured in the United States of America
Dover Publications, Inc.
180 Varick Street
New York, N.Y. 10014

Contents

CONTENTS

Foreword to the Dover Edition

Inflation and a rapidly changing world have not altered the basic aim of this book: to present in simplest form a practical guide for the beginning farmer, telling him the elementary things that the more specialized texts omit. It indicates where more technical information is available if and when needed, and sets forth the *easiest*, as well as the more perfectionist (and more strenuous) practices.

However, cost figures have varied widely in recent years, so prices that appear in the following pages should be translated into current local terms. Financial returns likewise vary because men, land and prices differ. Virtually nothing is said here, therefore, about possible profits because it would need to be so qualified as to be meaningless.

The additional readings at the end of each chapter are likewise dated. Many of the publications are now out of print; many contain information that has been superseded. The bibliography following offers an updated list.

Whether the beginner wants an acre or a full-scale farm, or chooses weekend farming or absentee ownership, information here should prove useful. But this book is merely a guide, and the reader's greatest fun and profit will be in developing his own methods, suited to his own soil.

In preparing this book originally, I was particularly indebted to my brother, Ralph Jacobs, practical farmer and long-time farm real estate agent; J. W. Clark, former Dane County (Wis.) agricultural agent, and, for the chapter on bees, Eldon Marple, Hayward, Wis. Last and possibly greatest, I owe much thanks for manifold suggestions, both in writing and in farming, to my farm-raised wife Katherine.

Bibliography

Chapter 1

Whether you should go to the country, how big a farm you need, and what to do when you get there have been discussed in dozens of books. Recent books emphasizing the practical aspects are: *Farming for Self-Sufficiency* (1973), by John and Sally Seymour, Schocken Books; *The Country Guide for City People* (1973), by Chase Collins, Stein and Day; *Homesteading* (1975), by Patricia Crawford, Macmillan; *Buying Country Land* (1973), by Eugene Boudreau, Macmillan; *Five Acres and Independence* (reprinted 1974), by M. G. Kains, Dover Publications; *Country Land and Its Uses* (1974), by Howard Orem, Healdsburg, Calif.; *The New Pioneers Handbook—"Getting Back to the Land in an Energy Scarce World"* (1975), by James Bohlen, Schocken Books. Among books dealing with the joys of country life is: *Country Matters* (1973), by Vance Bourjaily, Dial.

Chapter 4

The Complete Book of Home Preserving (1955), by Anne Seranne, Doubleday; *Making Butter on the Farm*, by E. S. Guthrie, N.Y. State College of Agriculture, Utica, N.Y. From U.S. Department of Agriculture: *Home Canning of Meat and Poultry* (1972) G 106; *Home Canning of Fruits and Vegetables* (1972) G 8; *Making Cottage Cheese at Home* (1975) G 129; *The House Fly—How to Control It*, L 390. From Superintendent of Documents, Washington, D.C. 20402: *Selecting and Raising Rabbits*, AB 358 (for sale only).

CHAPTER 8

From U.S. Department of Agriculture: *The Farm Beef Herd* (1975) F 2126; *Raising Livestock on the Small Farm* (1966) F 2224.

CHAPTER 9

From U.S. Department of Agriculture: *Hog Castration* (1960) L 473; *The Meat Type Hog* (1972) L 429; *Slaughtering, Cutting and Processing Pork on the Farm* (1973) F 2138.

CHAPTER 10

From U.S. Department of Agriculture: *Raising a Small Flock of Sheep* (1966) F 2222; *Housing and Equipment* (1969) F 2242. From Superintendent of Documents, Washington, D.C. 20402; *Docking, Castrating and Earmarking Lambs* (1969) No. 551 (for sale only).

CHAPTER 11

Starting Right with Poultry (1972), by G. T. Klein, Garden Way, Charlotte, N.C. From U.S. Department of Agriculture: *Poultry Management* (1972) F 2197; *Raising Ducks* (1960) F 2215.

CHAPTER 13

Wyman's Gardening Encyclopedia (1971), by Donald Wyman, Macmillan; *The Organic Gardener* (1972), by Catherine O. Foster, Knopf. From U.S. Department of Agriculture: *Insect Control Without Insecticides* (1975) G 211; *Selecting Fertilizers for Lawns and Gardens* (1973) G 59; *Growing Vegetables in Home Gardens* (1972) G 202; *Mulches for Your Garden* (1971) G 185.

CHAPTER 14

See material listed for Chapter 13.

CHAPTER 17

From U.S. Department of Agriculture: *Beekeeping for Beginners* (1975) G 158; Correspondence Courses: from Pennsylvania State University, Extension Division, University Park, Pa.: *Beekeeping*, Course 70. From Ohio State University, Columbus, O.: Course 16.

CHAPTER 18

A Sand County Almanac (1977), by Aldo Leopold, Ballantine; *Trees and Shrubs, Where to Buy Them*, Brooklyn Botanic Garden, 1000 Washington Ave., Brooklyn, N.Y. 11225. From U.S. Department of Agriculture: *Dwarf Fruit Trees, Selection and Care* (1976) L 407; *Managing the Farm Forest* (1971) F 2187.

CHAPTER 19

From U.S. Department of Agriculture: *Part-Time Farming* (1974) F 2178; *Leasing Contracts*, F 2164.

CHAPTER 23

Farming of Fish (1968), by C. F. Hickling, Pergamon Press; *The Maple Sugar Book* (1971), by Helen Nearing, Schocken. From U.S. Department of Agriculture: *Catfish Farming* (1975) F 2260; *Trout Farming Could Be Profitable to You* (1969) L 552; *Making Land Produce Useful Wildlife*, L 2035; *Ponds for Water Supply and Recreation* (1971) AH 387; *Building a Pond* (1973) F 2256; *Simple Plumbing Repairs for Home and Farmstead* (1972) F 2202; *Soils and Septic Tanks* (1971) AB 349; *Use of Concrete on the Farm* (1975) F 2203.

CHAPTER 25

From U.S. Department of Agriculture: *Records—Accounts* F 2167.

PART I

The Life You Can Lead

-I-

WHY ARE YOU REALLY GOING?

"How can a man take root and thrive without land?"
—JOHN BURROUGHS

THE question of why you are going to the country—the sort of life of your own that you want to lead—is basic. A little soul-searching now may save an expensive false start, with the consequent disappointments.

Maybe you just want to get away from city noises, airless apartments, the absence of greenery, and space. Perhaps you have growing children, and want to give them a chance to grow up in close touch with animals and woods and water, learning self-reliance, bodily skills, and understanding.

There are lots of other valid reasons for going to the country. Perhaps you have a taste for really fresh fruits and vegetables. You may have wanted for years just to "fool around with a few animals," or to supplement your income by raising most of your own food. It may be a matter of health that calls you outdoors. Maybe you just want to own a farm, without actually living on it.

More likely, a combination of reasons, including some which may not be set forth here, is egging you on. Whatever they are, they will be your signpost pointing down one of the avenues of agriculture. These avenues, to be sure, are not one-way streets, from which there is no return or cross-connection, but as in any journey, it is easier and quicker to take the right road the first time.

Are you, for instance, planning to make a clean break with city life, and take to full-time farming? If that is your purpose, you can locate in virtually any section of the country whose climate suits you, and where the type of agriculture that attracts you can be carried on. On the other hand, if you are going to hang onto that city pay check, but want several acres for subsistence farming, you will be looking for a place within driving distance of your work, or some place where similar jobs are available.

Or perhaps you just want "one acre and a view," near enough to your present job for year-round living; or possibly something farther away, as a summer and week-end retreat.

Whichever of these choices lies in your mind, you can still work out a satisfactory solution, if you take your time and know what you want. (See Chap. 20, "Choosing the Land.")

A romantic feeling that the country is an idyllic place to live in is not always sufficient reason for going there. Unless you can pin down your desires more closely, you are merely inviting trouble and expense. Neither is the idea that a full-scale farm, or even a small place in the country, is the highroad to financial ease. There's no magic about farming which can make an acre produce more for you than it will for the next man. Good farms in good hands, to be sure, make a comfortable living for the owners, but they reflect the application of real skills and knowledge to a setup capable of producing a profit. Even the best farmer won't make a go of it on an overbuilt or soil-depleted farm.

The same situation applies to the "one-acre" farmer or the man with a subsistence farm. If a hundred-acre dairy farm, for instance, produces a good living—but not much more—for a competent farmer, the man on three acres should not be disappointed if his farming returns are in proportion to his acreage. To be sure, by intensive production of truck crops he may show a much greater return per acre than the large-scale farmer, but he will be farming his back and arms, as well as the land. If he has the time, he can increase his income substantially on the three acres—but if he kept books on his hourly rate of pay, he might be disappointed.

All of which is just another way of saying that farming is a way of life, even on just a few acres of land, and he who lives that way does so because he enjoys it, rather than because he expects to pick up diamonds and rubies every time he stubs his toe.

ADDITIONAL READING MATERIAL: Whether you should go to the country, how big a farm you need, and what to do when you get there have been discussed in several good books. On size, for instance: *Ten Acres Enough*, by Edmund Morris, New York, 1865, and *Five Acres Too Much*, by Robert B. Roosevelt, New York, 1869, were echoed recently, with encyclopedic aspects, by *Five Acres and Independence*, by M. G. Kains, and *The Have-More Plan*, by Ed and Carolyn Robinson.

Recent books emphasizing the practical aspects of farming are: *A Practical Guide to Successful Farming*, (by 36 experts) edited by Wallace S. Moreland; *Success on the Small Farm*, by Haydn S. Pearson; *The Farm Primer, a Manual for Beginner and Part-Time Farmer*, by Walter M. Teller; *Land for the Family, a Guide to Country Living*, by Axel F. Gustafson; *Farming for Security*, by William B. Duryee; *Buy an Acre, America's Second Frontier*, by Paul Corey; *Farm for Fortune and Vice Versa, a Handbook for City Farmers*, by Ladd Haystead; *Three Acres and Liberty* (1907) and *A Little Land and A Living* (1908), both by Bolton Hall, and *Practical Farming for Beginners*, by Harold A. Highstone.

Among books dealing with the joys of country life are *Pleasant Valley* and *Malabar Farm*, by Louis Bromfield; *One Man's Meat*, by E. B. White; *Glory Hill Farm—100 Acres Farmed by an Amateur*, by Clifton Reynolds; *Country Colic*, by Robert Lawson; and *We Chose the Country*, by Herbert Jacobs.

*Pamphlets:** From Superintendent of Documents, Government Printing Office, Washington 25, D. C.: *Shall I Take Up Farming?* published 1945 by U.S. Dept. of Agriculture, 50 pp., ill., 15¢, Catalogue No. W 1.55:35. *Planning a Subsistence Homestead*, revised 1940, 20 pp., ill., Farmers Bulletin 1733, 5¢.

---

* Since this book went to press, prices of some of the pamphlets on farming, offered by the U. S. Government and listed here and elsewhere in this book as supplementary reading material, have been increased. It is advisable to check the new prices before ordering.

BEFORE YOU GO——

YOU ought to realize clearly that country life has a less inviting side. This chapter deals with the seamier aspects of farming, although personally I believe the good side far outweighs the bad.

Certain difficulties spring naturally to everyone's mind: the power line which breaks oftener and for longer periods in the country, stilling most of the machines with which you work; the danger of being snowbound; the daily round of hard physical work. But there are smaller troubles, which can add up to quite a bundle. A farm woman who had been raised in the city summed up her complaints this way to me:

"I can't just run around the corner and get a new pair of stockings when I need them. There's no chance for window shopping whenever I feel like it. If I want to take in a movie, it means a long drive there and back. I can't even run over to the drugstore without waiting for my husband to take me in the car, and I can't step around to the corner grocery if I need something right away for cooking, because there isn't any grocery nearer than five miles. Sometimes in winter we can't even get out if we want to for a couple of days, because the plows haven't opened the roads."

Each of these little irritations may seem trivial—but they can add up to cause for feminine revolt against country living, unless the advantages outweigh them, or unless special efforts are made to see that these wifely cravings are satisfied. Unless the female half of the partnership is going to make a successful effort to be contented in the country, you might just as well stay in town, because you'll be back there soon anyway.

The old-timers can truthfully point out that country living is no longer as inconvenient as it used to be—but nothing is static in this world, and while country life has improved in convenience, so has the city, so maybe we're just where we were before.

On the physical side, country living now is very close to that of the city in mechanical conveniences. Wherever the electric lines run, farmers can have running water in the house as well as the barn. Pressure water systems and septic tanks can make the frosty outside toilet a relic of the past. An electric stove—or, if the cost of electricity is too high, a "bottle gas" stove—takes the place of the old wood- or coal-burning range. Oil-burning space heaters or furnaces, or coal stokers keep all corners of the house warm.

Nevertheless, such conveniences are only good while they run. When the belt on the force pump gets loose or breaks, the water system stops functioning until you put on a new belt. There's no trick to doing it, but the effort is somewhat greater than calling up the city water department and registering a complaint. At the other end of the system, septic tanks have sometimes been known to get full and back up. A handy person can make a little arrangement of pipes with a centrifugal pump to take care of the situation, but if this is beyond him, he will have to hire a plumber.

Every improvement in facilities since the days of the log cabin, running stream, and open fireplace has meant more complicated machines. It is the regrettable tendency of machines to break down, wear out parts, or just simply refuse to function for various unfathomable reasons. A man who is handy with tools can make them go again, but occasionally he is likely to reflect sourly that most of his life is spent in repair work, rather than in productive endeavor.

Even with the best of machines, which are all in perfect order and run like clockwork, farm life still requires long stretches of good hard work. When the hay is ready to cut, you start early and work late, ignoring any preconceptions about the eight-hour day. Likewise in the spring, when the land at last is just right for plowing and seeding, you will be at it almost round the clock, to get ahead of the weather.

Sometimes, in fact, modern inventions seem an added refinement of cruelty to aching muscles. In this class the farmer is likely to consider lights for his tractor, so that he can go on pounding his tail on the tractor seat far into the night to get his plowing done. Such pounding is even apt to cause him to invest the next day in one of. the hydraulic seats which make tractor driving easier to endure.

For the man on one acre, the shift from city to country living is easy, because he can get by doing virtually nothing. The part-time or subsistence farmer will find the going tougher. He has more work, more machines, more worries. But the man who steps from a city job to full-time farming has the most difficult adjustment to make. Even if he has been engaged in hard labor, he will be using different sets of muscles on the farm, and his back will be weary many a night before he becomes hardened. And it will be weary many a night thereafter, too, because farm work is *hard*. His mental muscles will also get a workout, because modern farming has become a highly skilled occupation, requiring vast quantities of information and "know-how."

The city man intending to engage in full-time farming would be wise to try it for a year as a hired hand or tenant. In that way he will gain practical experience with all expenses paid, and he will rapidly discover whether he is really cut out to be a farmer after all. This is a good thing to find out before making a heavy investment in land, machinery and livestock.

-3-

WHERE THE FAMILY FITS IN

HERE and there, you will find an old widower grimly tilling his acres, but mostly, country life is for families—for the richest kind of family life. I do not think it is possible, without great effort, to achieve the same sort of thing in cities. There pets are limited to a dog on leash, a canary in a cage, an occasional cat. Recreation usually means something outside the home—the movies, the "character-building" organizations for children, the synthetic projects which have no real meaning or educational value, for even the child can see that they serve no useful purpose at the moment. Usually the emphasis seems to be on things which take members of the family their separate ways, outside the home, rather than bringing them together inside it.

Compare this meager fare of the city with the boundless possibilities of the country, and you can better understand why the country-bred child is usually self-reliant and able to land on his feet. By its very nature the family-sized farm is a family enterprise. I do not mean that it should become a family sweatshop, with children doggedly driven beyond their skill and strength until they flee to the city as from a chain gang. I do not hold that children have a "duty" to help support their parents before they are even able to support themselves. Let's put it a different way—the parents have a duty to give their children the experiences and training which will make them happy and useful members of whatever society they may later find themselves in.

9

You can't put a seven-year-old in a high-powered car and let him drive it out on the highway, but you can put that boy or girl on the front of a hayrack, where he will have the time of his life guiding a team around the field, while father (close behind him in case of emergency) piles the hay around the rack as it comes off the loader. At the first turn, in spite of copious advice, he'll nearly upset the wagon by turning too short, but before the hay is in off that field, he'll be a pretty fair driver, besides feeling that he is really a help to his daddy.

Some day, when you're going back to the field with an empty wagon, the boy will get his first chance to drive the tractor. Almost before you know it he will be handling the machine like a veteran, even though it may be years before he has the strength and skill for the heavier tasks like plowing and cultivating. So with all the other machines of the farm. Before he has even graduated from high school, he will have become expert in an occupation in which he can always earn a living; and that is one of the finest gifts a parent can make to his offspring.

Possibly nothing can be done prenatally, but surely in their first years children can begin to absorb a feeling for animals and for tasks done in common. Mine went out with me to the barn before they could walk. I carried them on my shoulders to the haymow and set them in a corner while I pitched down fodder. Down below, they were wedged between a couple of straw bales until they could walk, and as soon as they could toddle, screamed with delight at the chance of helping to grain the animals. Sometimes they dumped in too much feed, or got it in the wrong places—but I always remembered that the most worth-while crop any farmer can raise is the human one, so I let them make mistakes in order to let them learn.

Corn harvesting is another great occasion for children. To ride the wagon back of the corn-picking machine, and watch the yellow ears tumble out from the conveyor belt is satisfying, but I think every farmer owes it to himself and his children to leave at least a couple of rows standing, to be harvested by hand. If he has a horse instead of a tractor to pull the wagon, so much the better. (The horse will go ahead or stop at a shouted command, whereas the tractor requires someone to get up on the seat and drive it.) The children can husk as much as they please, search for red ears, measure to see who found the longest ear, ride the wagon, feed cornstalks to the horse, and watch for pheasants and rabbits. What matters if they don't spend every minute husking corn? If they did, you would be raising a generation of little old men.

One of the great experiences of a farm childhood is growing up with animals. Helping to teach a calf to drink from a pail, "bottling" a lamb which has lost its mother, studying the habits of old hens to see where they have stolen nests—these are among the experiences that will be remembered in later years. They don't have the strength or skill for it, but children love to feel that they are doing a man's work, or at least helping to do regular work. When I made a barn floor scraper of wood, I made two additional little ones, with short handles, and soon I found that the children kept the barn floor tolerably clean.

Children rebel at being told to go out alone and weed the garden, but *they will work alongside their parents*, thinning carrots, pulling out weeds in the row, and scratching with a cultivator hoe. They like a goal in the garden—"three more rows and we quit"—but they are quick to lose interest if the quota is raised when they are near the end. After all, the sense of doing a full day's work is something which comes only with maturity, but they can understand and take part in short spurts with a fixed goal in view.

I never made any distinction between boys and girls in teaching farm chores like milking, driving, and the care of animals. Both are the better for learning these tasks, though possibly the girls should merely learn how to do them, rather than be kept at them steadily. After all, they have kitchen skills to acquire, which they can't learn in the barn. Even though it is essentially the women's province, the whole family can pitch in during the canning season, picking and preserving fruits and vegetables, so that the next winter every member of the family can know just where the food came from, and how much work it took for the preservation.

Every boy and most girls seem to go through the stage of wanting some small animals as pets, wholly dependent on them for survival and care. On a farm, they can have them. Whether it is pigeons in an old applebox in the hayloft, rabbits in a hutch near the chicken coop, or a setting of duck eggs under a mother hen, a farm has the space, the feed and usually the housing—without fear of zoning laws or complaints of neighbors. Usually, after the novelty wears off, they need adult supervision to see that they actually do care for their young charges instead of neglecting them, but along the way they acquire some understanding of what it means to stick to a job—and possibly the folly of getting involved in something they are not really interested in.

This is the age, also, for the bold huntsman, for nailing skins of wild beasts to the barn door, even if the animal is only a gopher. Every family will have its own idea of just when a boy is to be trusted with a gun, but he can get nearly as much fun out of a bow and arrow, and be considerably less destructive. Setting trap lines in the winter, fishing in summer, and hunting both fall and winter, can be good teachers and good health builders.

Beginning around the age of ten, boys and girls can take part in one of the most worth-while rural endeavors—the animal, crop, and homemaking projects of such organizations as the Four-H Clubs. In many counties of the United States children now have their own Four-H or junior fair, which often takes the place of any other fair in the county. Under supervision of the Four-H Club leaders, each child picks a project and tries to carry it through. In a calf project, for instance, a boy or girl acquires a calf by purchase or gift, and is entirely responsible for feeding and training the animal. The young agriculturalist works out a feeding formula, and begins early to train his young charge for the show ring. He walks him at the end of a halter, teaches him to stand in position to catch the judge's eye, gets the calf used to being among people instead of in a lonely pen in the barn.

At the county junior fair the young exhibitor bumps into competition for the first time, and begins to learn many salutary lessons.

He discovers to his surprise that a lot of other boys and girls have been raising calves, and that some look better than his. He learns about grooming his calf, and when he gets into the judging ring, the other boys and girls teach him some of the elements of show-manship. For instance, there's the trick of diagnosing where the judging line will start, and getting your animal to that spot first, so that the judge will have your calf firmly in mind, and so that the calf himself will be rested and calm, rather than upset by a last-minute entrance. Youngsters learn to take it with a smile when the judge motions their prize animal out of the ring as not being up to the class of the few best animals. The young exhibitor may be biting his lips, but he is acquiring the knowledge of how to accept defeat—and come back fighting for the next encounter. After his first defeat he is more likely to listen to the advice of others—and there will be plenty of it—on how to do the best job of fitting and exhibiting.

For those who do well in the county junior fair, there is always the possibility of the stiffer competition in the state fair, and the regional junior livestock shows. If a boy or girl is fortunate enough to start with a first-class animal, and develop it to perfection, the heights of success beckon with really worth-while rewards. In addition to the coveted ribbon and a picture in the paper, he can count on awards which may run into several hundred dollars. Even the dizzy pinnacle of a grand championship at the international livestock expositions in Chicago or Kansas City is not beyond him, for these honors have been won several times by farm youths, com-peting against veteran stock breeders.

Animal projects are especially suitable for growing children be-cause the animals develop so fast that there is constant change, while the cash result at the end of the project usually brings in a good sum.

And where is mother in all this? Naturally, she is busy helping her daughters to become skilled homemakers, but she also plays a con-siderable part in the general farm life. After a family conference on whether to call the veterinarian for an ailing cow, mother may be pressed into service as assistant or even chief operator, if medicine is to be given. Many a farm wife has left a hot kitchen to engage in the even hotter job of loading a steer or a couple of pigs onto a trailer. If she has an egg route in town, she and her husband may spend a quiet evening catching up on each other's ideas while they clean eggs together against the morrow, or they may be killing and plucking a dozen chickens together, to sell along the route. Whatever the triumph or disaster, the female member of the partnership will certainly be called on to share it—to double the joy of a good stroke, or halve the sorrow of a sheer miss.

Unlike in the city, where the wife and children may have only a vague idea of what the breadwinner does to earn his pay check, the farm family is keenly aware of all that is going on for the family's betterment. In fact, there are no secrets as to whether the family farm enterprise is doing well or ill, because all can count the cows, see the number of milk cans set out for the hauler, measure the progress of plowing, cultivation or harvest, watch and number the growing pigs as they fatten.

It is difficult to estimate the effect of farm life on the family's spiritual values, but some straws show in the wind. Church, for instance, means more in the country. After a week's comparative isolation on individual farms, people get together on Sunday for sociability as well as religion, and the church becomes a dominating community center.

Whether the school adds to the family's spiritual resources depends somewhat on the school, but events like the annual Christmas program and the school picnic are often geared for adults as well as children, and thus serve to draw the neighborhood closer together.

The farm certainly helps supply that great need of childhood— the feeling of security. Instead of the uncertainties of a rented dwelling and a father absent most of the day, the country child can see for himself that his home is thriving and reasonably permanent, while the day-long presence of both parents is further testimony that his small world is fixed and immovable.

ADDITIONAL READING MATERIAL: *Home Tanning of Leather and Small Fur Skins,* published 1923, revised 1947, F 1334 (free from U.S. Dept. of Agriculture).

-4-

ABUNDANCE CAN BE YOURS

LIFE on the farm can bring you almost every delicacy, as well as more substantial fare.

Naturally you have, or soon will have, a quick freeze locker. Surprisingly, it need not be a big one. What you need, to make the most of your opportunities, is from five to ten feet of cold storage space in your own home to insure variety always available. "But I can't possibly get all I need for a year in just ten cubic feet of storage space," you complain—and rightly. You can get more space in two ways. One way is to buy an extra large freezer—say eighteen to thirty cubic feet. At present prices, these run into a lot of money. The more economical way is to rent frozen storage space in some near-by locker plant. You can rent five cubic feet of storage space for between $10 and $15 a year, depending on the neighborhood, but if you buy a freezer to obtain this same space in your own home, it will cost you anywhere from $25 to $35 a year, counting investment, depreciation, repairs, and electric current. The locker plant will also butcher and process your animals.

It is usually possible to store several boxes of overflow from your own freezer in a locker plant for a very small charge. Remember that the secret of the most economical use of a locker is to fill and empty it at least four times a year. Your big push for storage space will come toward the end of the summer, when you have put away scores of quarts of vegetables and fruits. Keep enough to last you three months, and send the rest to the locker plant for storage until winter. Time your meat butchering so as to avoid the peaks of the vegetable season. With fifteen cubic feet of storage space (say ten at home and five rented in a locker plant), you can handle a whole steer. The time to butcher him is early spring, so that you will be eating the meat, and thus making more space, as the summer vegetables and fruits come on. By midsummer you can slip in a lamb or two, and in late fall a hog. By the middle of winter—assuming your family is good-sized—you'll be ready for another steer.

Before the advent of the quick freeze locker, the country menu left much to be desired. Fresh meat was plentiful when a steer or hog was killed, but after that it was a wearying succession of canned or smoked meat. Vegetables were always canned, by the hundreds of quarts. The diet was wholesome and nourishing, all right, but it certainly was lacking in appetizing variety. The farm wife nearly killed herself off getting all the meat canned, and her family got fed —and fed up—the rest of the year on the pallid chunks of meat which were their lot.

Those days are gone, on most farms. Because freezer space is still expensive, many things are still canned, but meat is not among them. Steaks such as you can rarely obtain in the most expensive restaurant are yours the year round; asparagus and strawberries frozen to last through the fall and up to Thanksgiving; half a dozen chickens and ducks, killed at their prime; corn on the cob, broccoli, green beans —the list is endless, and the table results are delectable. Any person with imagination can see the possibilities, including pheasants and other wild game, fish, and the vast variety of pies, cakes, fancy breads and other foods which can be prepared ahead of time and fast-frozen for use later. Even the children's school lunches can be prepared weeks ahead of time. They will thaw out just right, between the time of going to school and the lunch hour. Dozens of books and pamphlets are now available, telling you how to get the most out of a quick freeze locker. Very few, however, stress the expense of a locker, and the possibility of cutting corners by renting extra space for a few months when it is needed.

Nevertheless, to get the most out of frozen foods, you should

have at least a small unit in your own farmhouse. One reason is to
have a variety of meats and vegetables always available, without
a trip to the locker plant every other day. The other reason is that
you will not put up so much if you have to haul it to the locker
plant to be processed and frozen. It just isn't worth while to dash off
to the village with two or three quarts of strawberries or beans, but
if you have your own quick freeze, you can preserve small extra
quantities every day, and move them later to surplus storage space
if you need to.

The quick freeze locker offers the most spectacular evidence of
the variety of foods now available to the farmer, but it is only one
aspect of a tremendous cornucopia. Even the rich man on his
thousand-acre estate, complete with gardener, yardman, herdsman,
butler, and three cooks, will not be able to enjoy the bounty that is
at your hand, for the ponderous machinery of such a large institution makes it practically certain that his vegetables will be picked too
early in the day to come really fresh to his table.

Did you know, for instance, that the sugars in sweet corn, which
give this delicacy its name, change to starches within three hours
after the corn is picked, and that the real connoisseurs of corn insist
on no more than ten minutes between the time it is picked and
when it is in the kettle of boiling water? After you have once tasted
corn really fresh from the field, you will wonder why anyone buys
the starchy, tough cobs which have sat in the grocer's bin for as
long as three or four days. Not all vegetables are as sensitive as corn
to the time lag between harvest and sale, but surely none is improved
by the process. Fresh beet greens, with inch-sized beets tagging at
the ends, the leaves still too fresh to have wilted; small carrots,
thinned from the long rows which will make your winter supply;
freshly podded peas or lima beans—these are the things which money
literally cannot buy, though work can.

Not only the garden brings things fresh to the table. Trees and
the roadside also contribute. In early spring the children dash out
along the fencerows to gather asparagus before the "city slickers"
can arrive in their cars to purloin it—and it is hard to determine

which is more attractive, the full pail of tender asparagus or the beamish faces of children full of a sense of accomplishment. Later, depending on the locality, come wild strawberries, black raspberries left by the cows along the line fences, cherries for pie, from your own tree or picked on shares at the neighbor's, blueberries, red raspberries, and blackberries, the first yellow transparent apples, ready for pie or apple sauce. Then as fall comes on, follow wild plums for jelly and jam, wild grapes from the old fence near the meadow, wild crabs with a honey-sweet and sticky bloom upon them, ready to be tinted with a bit of grape for jelly or left in their own pink lusciousness.

With the quick freeze full, and the fruit cellar shelves crammed with hundreds of quarts of tomatoes, beans, beets, and other vegetables and fruits, you are ready for the fall harvest which will help to see you through the winter. Late cabbages to be tied in newspapers and hung from the ceiling; deep green or bluish Hubbard squashes, good until February at least; nets of onions, the white-fleshed bland Berumudas nestling beside the more astringent seed onions which have come on to plumpness since April, and the tiny Barlettas, pickling onions ready to be made into a Thanksgiving savory relish. On the floor, in a cooler corner, bags of early potatoes, red or white, ready to be used up before the late potatoes, and barrels of apples, both the cooking kind and those rosy-skinned ones ready to be slipped into a child's lunchbox. In sand near by, carrots, beets and celery, in whatever quantity you can put away.

Surely this is no definitive list, and every family will want to add and subtract, according to its own preferences. But dozens of other things crowd for attention. Have you left those apply tree prunings in a convenient fence corner, ready to make an outdoor steak-fry fire for a festive picnic?

Did you plant popcorn this year—at least forty feet from any other corn, to prevent cross-pollination—and have you hung the ears back of the stove, to dry out for parties for the young and old folks? Were there a few hills of sorghum or sugar cane for the kids to chew on? And sunflowers in a sheltered spot in view of the house, to attract songbirds in fall and winter?

If the physical larder is well filled, the spiritual one should be no less replete. There are winter nights, as you come back from the barn to the house, when Orion flames across the sky and the North Star dips to the nearest hill. Then is the time to teach the children the constellations, just as the middle of August, when the earth's orbit crosses the Perseids, is a time to get the children out into the open, away from the trees, to show them the majesty of meteors— the lesson that they live not only in one world but in one impinging universe.

Perhaps the greatest discovery on moving to the country is that of the horizon. Sometimes in cities, from the ironclad pinnacle of an apartment, the horizon can be seen, as through a haze, darkly, and a sunset bonfire of the east-west streets tell of distances, but mostly it is your neighbor's wall and your oppressor's factory that loom before you. In the country, it can be a ridge of poplars three miles away, with the setting sun lacing the snow-clad fields with red between them, or the long shadows, in summer, of willows by the roadside darkening a field of grain. Above them, clouds—great cumulus towers of snowy white or pink, mare's tails high above, presaging an impending storm; the dark, ragged edge of a line storm sweeping overhead; or the little wispy clouds scudding below a storm wrack. And sometimes, nothing but blue overhead and on all sides—as far as the eye can see.

ADDITIONAL READING MATERIAL: *Making Butter on the Farm*, E. S. Guthrie, N.Y. State College of Agriculture, Utica, N.Y.

From Superintendent of Documents, Washington 25, D.C.: *Making American Cheese on the Farm for Home Consumption*, 18 pp., Farm Bulletin 1734, 5¢; *Making and Storing Farm Butter for Home Use*, 16 pp., ill., 10¢, Cat. No. A 1.9:1979; *Home Canning of Fruits and Vegetables*, 24 pp., ill., 5¢, Cat. No. A 1.64:64; *Home Canning of Meat*, 16 pp., ill., 5¢, Cat. No. A 1.59:110; *Beef on the Farm, Slaughtering, Cutting, Curing*, 34 pp., ill., 5¢, Cat. No. A 1.9:1415; *Pork on the Farm, Killing, Curing and Canning*, 38 pp., ill., 10¢, Cat. No. A 1.9:1186.

-5-

PLAN TO BE A SPECIALIST

FARMING is a way of life—but there are literally scores of ways to live it. I conceive of this way of life as a sort of dual thread: you should be famous for something, and you should not spread your time and energy on so many projects that you do none of them well, and have no time to enjoy life.

It doesn't really matter too much what you are famous for, but be a specialist in something. You can set your sights high, and try for the grand champion dairy cow of the United States—or you can just run a good hotbed and have extra plants for your neighbors for their gardens. Your fame will be about the same in either case. One will be local, and the other national. You may get more satisfaction out of the local notoriety, for it is no mean feat to produce plenty of tomato, celery, broccoli and cabbage plants at a time when they will do the most good in the ground. On the other hand, raising the champion dairy cow may mean the much less complicated process of laying out a big wad of cash for a good heifer from a championship line, and then hoping for the best.

In between the champion cow and the champion hotbed are hundreds of possibilities: you can raise certified seed for oats or wheat, you can produce good stands of exceptional hay, or make good hayseed. One possibility, in a dairy region, is to maintain a beef bull like an Aberdeen Angus, for loan to the neighbors. It can be bantam chickens, bronze turkeys, Yorkshire hogs, Ladino clover seed, or almost anything you can think of. The important point is to pick out a possibility that is within the capabilities of your land and yourself, and develop it as the particular specialty of your own farm.

There is a certain prestige value in such specialization—and cash profits too. If you do one thing well, much will be forgiven you that is merely done indifferently. And as soon as you do one thing close to perfection, the others will be easier. With the uncertainties of weather, soil and seed, farming can contain many a disappointment. You can have a wonderful stand of hay early in May, and see it all blasted before harvest, by drought or grasshoppers. In the same way, disease or accident may carry off your prize animals. The specialist, however, can usually counterbalance nature and accident after he has been at it a few years. If you are specializing in sweet corn, for instance, you learn to plant when the ground is in perfect tilth, and you may plant as many as ten plantings, being certain that at least half of them will come through regardless of too much or too little rain.

We all chase the bubble of success, in one way or another, and specialization is a good way of achieving the biggest bubble. To be known, even in a limited region, as supreme in at least one field, does things to your ego which make the labor and worries of farming seem trivial.

If you want the easiest kind of success, try raising jumbo squash or pumpkins. If you've got a good vine, with some fruit set, pinch off all the other blossoms, and break the ends of the runners, so that the energy of the plant is not dissipated. About two feet toward the root from your biggest squash or pumpkin, make a slit in the vine, put in a cloth wick, and keep the wick soaked in a pan of sugar and sour milk. That squash or pumpkin will zoom to truly astronomical proportions, and can reach as high as two hundred pounds in weight, compared to fifty pounds for a normal squash. Placed in the window of the local store, with a card telling who raised it, it will establish your fame forever. (However, after the exhibition, don't let anybody try to cut and eat the squash. It may be woody and unpalatable.)

One advantage of specializing is that it concentrates your energies, and keeps you from scattering them over too great a variety of endeavors. Remember that each new kind of animal you introduce to your farm usually means a new type of building, and a lot of extra work at chore time. For instance, if you go in for raising riding horses because you happen to have a good horse barn on the farm, you will soon be laying out money for the traditional paddock of white-painted fencing, plus considerable sums for harness and gear.

Nevertheless, if you follow the cult of the horse, don't let anything set forth here discourage you from pursuing that bent as far as your energy and cash permit. Horses are among the most sensitive and interesting of farm animals. They have personality and behavior problems fully equal to anything the so-called human race can conjure up. I remember a friend on a farm west of me, who thought it would be a labor-saving device to train his horses to open the gates for him. He began by unlatching the gates, and taught the team to lean against the gate, and thus swing it open. He ended by finding the horses leaning against any piece of wood they came across, and a horse's weight being hefty, he was constantly meeting his horses in the farmyard or around the house when he thought they were out in the field.

Incidentally, there are two schools of thought about how to catch horses in an open field—assuming that they won't just let you walk up and bridle them, which most of them won't. One method is to attempt to drive them into a lane, where they are close enough to seize. You can try your voice, or sticks and stones—or even firecrackers, as one neighbor of mine used to do—to get them into the lane. Sometimes a pail of oats will do it, if the horses are of a trusting nature.

The other method is to get into your car and chase the horse around the field, honking the horn, and otherwise matching the forty or more horses under the hood with the one elusive equine in the pasture. If you are persistent enough, you may eventually capture the horse. Whether you win or lose, you will at least gain a new insight into the vitality of the animal kingdom.

If you have the equipment and fencing, and the strength of character to resist buying all the pacers and trotters that come along, horses can add to your income. For instance, if your farm is not too far away, some city horse lovers will ask you to stable their prize mares the year round. Before you settle on a price, however, inquire around at near-by stables. You will be surprised at how much people are paying to board a horse, and your own charges should reflect this weakness for horseflesh. Other possibilities are to keep ponies, hinnies or burros for your friends, either at a fee, or for your use of the animals.

On smaller acreages, and even on large farms, there is the possibility of specializing in little, rather than in big, things. You can raise fancy strawberry plants, herbs, or "whips" of fruit trees and shrubs. You can start your children on small animals like hamsters, or on pheasants or bantams, and develop a mail order business.

In short, regardless of the size of your farm, there is some specialty within reach of your capabilities. For your own satisfaction, and to make a name for yourself, don't fail to pursue it.

-6-

KEEPING YOUR BALANCE

ONE of the most impressive—and to me disheartening—sights as I drive into the city of a morning is to see the barrels and boxes of trash lining the curb, waiting to be picked up by the truck and carted to the municipal dump. It is a striking lesson in the difference between farm and city life, for on the farm nothing is wasted, and no one calls on a tax-supported body to remove things which he can dispose of himself. On the farm, things are in balance, and what you cannot use in one place finds a function in another location.

Perhaps trash disposal is a poor illustration of what I mean, but let us use it nevertheless. To be sure, many a farm landscape is disgraced by a rowdy collection of old tin cans, ashes and bottles near the back door, but it need not be so. Virtually every farm has an old gully, gravel pit or sinkhole which can stand filling. As the trash accumulates, the hydraulic lift on the tractor can cover it with fresh dirt, and gradually the field is extended to increase the cultivated area, and hence the profits. Sometimes it is a narrow section of driveway which can be widened, or a low spot in the barnyard, but even trash finds its uses on the farm. For instance, rolls of old wire from fences can be thrown into a gully to act as a partial dam.

The dairy farm is a perfect example of balance as applied to farming. If you want cream for butter or the table, the pigs and chickens will gobble up the skim milk. Manure from the barn and barnyard goes to the garden and fields. The extra grain from the mangers, which the cows will not eat, can be added to the ration for the pigs, who will also eat up all the garbage from the kitchen.

Old broccoli and cabbage stalks from the garden are hung up in the chicken house to give the fowls exercise and a bit of green to peck. The chaff on the haymow floor and from the spot where the straw bales were stored makes excellent litter for baby chicks and in the henhouse.

Whether you are a compost enthusiast or not seems to depend on what pamphlets you have read, but there are varying degrees of composting activities, to suit the temperature of your enthusiasm. At the head of the list are the devotees who build or buy special boxes for making compost, and who tend them with the zeal of a convert before an altar. These boxes permit you to dump in lawn clippings, weeds, garbage, and everything else organic at the top, while from the bottom—after an appropriate interval—you extract the rich humus which lightens and fortifies the soil for growth.

Those who don't want to go to the extreme of a box can make compost piles to use up the old weeds, degenerated hay, and practically everything else except minerals like glass and old iron. A compost pile is not just a heap of refuse. It is a scientific blending of organic matter in which heat, chemicals and moisture all play parts to produce the special soil-building material known as humus.

The four ingredients are: first, decaying organic matter, such as weeds, old straw or hay, lawn clippings, vegetable tops, garbage and the like; second, protein, in the form of manure, fish scraps, or table scraps; third, a chemical, such as lime or the phosphate of wood ashes; and fourth, soil, to furnish the leaven or nucleus of bacteria which will break down all the other ingredients into humus. Addition of ammonium nitrate speeds the process.

Because better and quicker results are obtained when air circulates through the compost heap, it is advisable to start first with a layer of branches, to let the air in underneath. For ease in handling, make the pile not wider than four feet, but it can be as long as you wish. On top of the branches pile the decaying organic matter, tramped down so that it is not higher than eight or ten inches. Next follow a few inches of manure, a sprinkling of lime or wood ashes, and a topping of two or three inches of soil. Using this layer form, you can build the compost pile as high as you like, taking care that the sides are a little higher than the center. To promote rapid decomposition, each succeeding set of layers should be no thicker than the first one. If the pile is higher than two feet, it should have air holes. You can produce these by putting a couple of two-by-fours lengthwise and crosswise in the pile, pulling them out after it is completed. It is also a good idea to have a couple of vertical air holes, which can be obtained in the same way. Taper the stack toward the top, and plaster the sides with dirt. The pile should be kept moist, but not soggy, to help the bacteria permeate it. Within a few days the compost pile will begin to generate heat—becoming almost too hot to stick your hand in it—as the bacteria in the soil begin to work. After about six weeks, the pile can be turned inside out, by forking or spading it over. The inside should be well decomposed by this time, but the outer edges will not be much affected, and they should be forked to the inside. In another month or six weeks the pile should have turned into humus, ready for working into the topsoil of the garden or the fields. It can be spaded in between the rows, or spread on top and worked in as a top dressing, depending on the season of the year.

If you don't like the job of turning the compost pile inside out, you can just let the heap remain a longer time, and use it as it is. The composting process works rapidly during spring and summer, but during the winter bacterial activity practically ceases, so you cannot start a compost pile late in the fall and expect to have it ready to use on the garden early in the spring.

The slowest and least satisfactory method of composting is just to throw the decayed vegetable matter into a pile and let it rot. The trouble is that the pile heats too much in the center and "burns out," rather than developing a humus rich in myriads of bacteria. Nevertheless, an old straw- or haystack which has sunk to within a few feet of the ground makes a valuable addition to the soil. Children play an unwitting part in helping the decomposition of such stacks. When they climb up to slide down the sides, they break down the weather seal at the top, so that water soon enters and assists in the composting process. An active child who loves to slide can make an old haystack fit for use as compost in about two years less time than if nature had been allowed to take her nonsliding course. How you can train children to slide on the stacks that you want for compost, and leave alone the ones you want for hay, is something that I will leave to the family psychologist.

Many other aspects of farm life lend themselves to this balancing use. For instance, calves or sheep will keep down the grass in the orchard and clean up the windfalls, while they fatten themselves. Old boards from a building you have torn down can be converted into field mangers for the cattle, or repairs for fences. The water from the milk cooler, run into a shallow depression, makes a fine duckpond. If you raise mink, or have a neighbor who does, the ducks themselves can form a happy combination with their fur-bearing enemies. Let the ducks run in the yard enclosing the mink cages, and they will grow fat gleaning the bits of food that drop through the cage netting.

Guinea hens, if you can stand their plaintive call, will pretty much feed themselves during every season but winter, as they roam the farmyard and fencerows with their young, and they make an interesting variation from chicken on the table. In a grassy meadow near a stream or pond, geese will maintain themselves completely.

Even a sinkhole or low spot on the farm is capable of conversion into something useful—namely, a fishpond. An acre of pond, the experts say, will produce as much food—to say nothing of sport—as an acre of land put into crops. With a bulldozer, or a scraper back of the tractor, you can make your own dam, enrich the water with commercial fertilizer, and stock it with fish from the state conservation department.

Woodlot and fencerow help to contribute to the balance on the farm. When the fence posts finally rot off at the ground level, they are still long enough for making electric fence, or they can be sawed in two and used for the fireplace. New fence posts will come from the woodlot, as a by-product when you are getting out fireplace wood or boards for a new hoghouse. Ashes from the brush you burn can be spread on the fields or garden, to add potash. (See Chap. 18, "Orchard, Woodlot and Fencerows.")

Even the man on an acre or two will find grass enough along the fences to tether a sheep, and the family garbage will just about maintain a hog. For the man on a larger place, the excess garden products which are overripe or too small to be worth selling can be fed to the livestock.

Curiously, an old "balancer" of our forefathers is now coming back into favor among pig raisers. Whey used to return in the cans when the farmer hauled his milk to the creamery, but lately, with milk sanitation regulations and difficulties of trucking to a distant milk plant, it has often been allowed to go to waste. Some farmers now are not only using up all their own waste products on the farm, but are finding it worth while to haul whey from the milk processing factory, rather than buying expensive mineral concentrates for their pigs.

And speaking of usability and balance, every farmer owes an eternal debt of gratitude to the inventor of the gunnysack. Surely no single piece of equipment passes through so many useful stages. Fresh and shiny, it arrives from the feed store, inclosing poultry or dairy feed supplements. Emptied, it can be used repeatedly for bagging up the farmer's own grains to haul to mill or field.

When the inevitable rips and small holes begin to appear, it can still be used for bagging up chaff in the haymow, to use as litter for the chickens. It is possible to prolong the cycle somewhat either by sewing patches on the small holes, or by the simpler method of gathering the edges of the hole together and tying them with a piece of twine.

As the holes get too big to repair, the gunnysack passes into new spheres of usefulness. It may still be sound enough to hold a few chickens or a young shoat on a short haul—though in neither case will it ever smell the same afterward. It can be tacked, double, over a broken window in the barn or henhouse, laid moistened over a box of vegetables stored in the root cellar, hung as a screen in front of a pig or chicken brooder, used to wrap a collection of greasy machine parts on their way to the blacksmith shop or garage, grabbed up to clean grease from machinery, laid down to kneel on when tending a sick cow.

But the gunnysack's life cycle is still not exhausted. Folded several times, it makes easier riding on the tractor seat; it can be thrown under the car wheels when you want to get out of the mud or off the ice; and finally, in its old age, lies down in front of the farmhouse door, to serve as a foot scraper. At last, matted and waterlogged, it goes out with the manure, back to the soil from which it came, to contribute its fibers to the soil as humus.

This survey of the gunnysack's chameleon life just scratches the surface of its usefulness, but it serves to indicate the infinite possibilities.

Besides these minor variants of the balancing theme, there is of course the much larger one of balance in the whole operation. The cycle revolves best around livestock. If you have enough livestock, they will produce sufficient manure for about a fourth of your farmland each year. With a four-year crop rotation, this will insure that your fields remain fertile, for the manure will increase the effectiveness of whatever commercial fertilizer you use. Without the manure it is difficult to maintain proper fertility balance on the farm, for the amino acids and other chemicals in the manure—not all of them fully understood as yet—seem to do things to enrich the soil which commercial fertilizers are not yet able to duplicate. Other balancers come during the crop rotation, when the nutrients taken out by greedy corn are restored by the nitrogen-accumulating legumes of the hay crop. (See Chap. 15, "The Living Soil.")

The pinch of financial necessity helps to keep the farm in balance. If you have too much livestock for the land to support, you will be buying extra feed until you get back into equilibrium. On the other hand, idle land or excess field crops will be an open invitation to bring the livestock up to capacity, which means full earning power.

In the broad sense, balance means the sort of crop rotation best suited to your particular soil and farm operation, enough livestock to consume your field crops and revitalize your fields with their manure—and enough, but not too much, work to keep you happily busy.

AROUND THE YEAR

WHEN you were a child, and spent two weeks with grandma on the old farm, you wished that you could spend a year there. Do you wonder now that you are grown what it would be like? Here follows a sampling of what could take place in my own neighborhood in the Middle West, but much of it applies to a good deal of the northern section of this country. I make no claim that this is a typical year. Let us call it, rather, an ideal year, when every farming practice works out just right, the weather is perfect for each stage of nature, and all your projects come to fruit.

JANUARY-FEBRUARY

I am an impatient soul who likes to advance the spring. Early in January I sneak out to the lilac bushes, one of the few shrubs which never become wholly dormant, and cut a few branches to put in water in the living room. In two weeks they will begin to put forth shoots, bringing greenery in the midst of snow. In a month, to be sure, the ends will begin to turn brown and they must be thrown out, but I have thoughtfully provided replacements, at two-week intervals, so the succession of green stays to delight the eye. In early February, the small wild plums are ready for cutting, and will show blossoms if the branches are kept in a dark place a few days before being put in water. At the end of February, branches of crab, for the delicate pink of the petals later, and a few sprigs of apple.

With the esthetic side assured, the more prosaic functions can be attended to. January is the traditional time for repairing machinery. The machine shed is cold, the dim winter light atrocious without the added assistance of electricity, bolts and gears are frozen in their accumulation of grease, and the whole atmosphere is wrong for tinkering with machines; but if you don't do it now, the press of other work later will prevent you. Broken hoes on the corn culti-vator are replaced; the mower blade is sharpened; a new canvas for the seed fiddler, fresh grease for all the wheels which will turn the summer through, and new couplings to match the new tractor hitch are provided.

Although the old saw says, "Prune when the knife is sharp," now is a good time to prune the orchard. Prune heavily, cutting back the new growth, opening the center of the tree to light, removing suckers at the base of the trunk, and branches that may rub and rot. There is a school of thought which holds that no man should prune his own orchard, because he cannot bear to be sufficiently ruthless with his own trees. I had an uncle with a twenty-acre orchard who would not even go near while his neighbor was pruning it. He was afraid he might cry, "Stop!"

Before the snow gets too deep, go into the woodlot and cut out the dead and dying trees, thin out the crooked and the crowded ones for fence posts, rails and firewood. Wedges will split the frozen eight-inch oaks into quarters for fence posts, and they can be sharpened for easy driving in the spring. If you have a neighbor with one of the new portable chain saws, you can cut up a lot of timber in a week.

By the end of February, if the sun brings a thaw during the day, and the night temperature does not fall below 25, the sap will rise in the maple trees, and you can tap them for your own maple syrup.

Some sheepmen don't like to have the lambs arrive until late April, but I like mine at the end of February, perhaps as another way of advancing the spring. Their sides puffed out like pillows, the ewes are brought into the barn, to have a dry, warm place for lambing. Their cries in the early morning are an unaccustomed and welcome sound as you open the barn door.

MARCH-APRIL

By the first of March, the last of the big blue Hubbard squashes will be passing across the dining table, apples will be low in the barrel, and the carrots either gone or too withered to be of use. It is an in-between time, with raw days outdoors and impatience in the house, particularly among the youngsters. You will be shoveling the snow and ice crystals from the entrances and exits of the culverts in the driveway, and clearing spillways to carry off the melting snows.

By the middle of March you can get out a big flowerpot and sow it thickly with celery seed, to put in the kitchen window, for transplanting later. You should have cleaned the old dirt and manure from the hotbed before frost last fall, but if you didn't, a crowbar and pick will help you now. Filled with fresh manure, well tramped down, and with dirt saved in the cellar last fall or laboriously dug from the field this spring, the hotbed is soon steaming the underside of the glass frames on sunny days, and you can plant your early cabbage, cauliflower and tomato seeds, leaving at least one row for a few tasty radishes and a handful of lettuce. Old burlap bags, weighted with a couple of stones, make a cover for the hotbed on cold nights.

You will be hauling manure earlier to the fields, while they are still white with frost, and before the sun has melted the top three inches of soil. When the frost finally goes out of the ground, you may have a few days of soggy driveway, when you will take the milk cans to the edge of the road rather than risk miring the truck of the hauler. Grass in the lawn and beside the driveway is definitely green instead of brown, and the blackness of the plowed land is turning to a light tan as the fields dry out.

With the sap rising, now is the time to show your skill to your children, by making a willow whistle. The art is almost forgotten, but you can learn quickly to ring the bark of a budless stretch, tap with the back of your knife to loosen it, and then slip off the shell, to deepen the notch and cut a slot from mouthpiece to tone chamber. You don't need to worry about being plagued by noise. The children will have so much fun taking the bark shell off and putting it back that it will soon break, but they will have learned a lesson in how the pioneers made the best of what they had.

After the frost is out and the ground begins to dry, tour the pasture and hay-seeded fields, to see whether the spring succession of freeze and thaw has broken the roots. If the root crowns seem to be heaved out of the ground, a tug at a few of them will soon tell you whether you will be planting emergency forage. There is no proof quite so convincing, that nature does not always treat the farmer kindly, as to pull out four inches of root.

Depending on the dryness of the ground, you can scratch up a few square feet of soil late in March or early April, for a "first-early" garden planting. Radishes, lettuce, carrots sowed in a wide row—even peas, if you have room for them—can be put into the ground. Frost and hail may get them all, but you have a better than even chance of having a few tidbits from the soil before the regular garden begins to produce. It's a chance worth taking. Last year's parsnips, dug as soon as the ground is loose, and fried in butter, have a peculiarly satisfying taste of spring, and should be eaten soon, before their tops begin to sprout.

Even before the grass is green, the chives in the perennial garden will be bristling with tiny green shoots—the very first new growth of spring to come from the soil to your table—mixed with hamburger, they will give an added fillip to this farm staple.

MAY-JUNE

By the first of May the cows are wild to get out on pasture, but the grass is still too short to be cropped safely. You measure the remaining hay in the loft, and wonder if you will have enough to spring out. If worst comes to worst, you can buy a few extra bales, or send the children to pasture the cows along the roadside. While they are doing it, they can take a pail and bring back the asparagus growing wild in the fencerows. An old bushel basket placed upside down over the rhubarb will produce high stalks in a week.

Field work, at the moment, is at a standstill. Oats were "mudded in" on the old corn stubble as soon as the ground could be disked in April, and are already mantling the brown soil with spears of green. It is time for fence building and repair. The ground is still soggy, for easy driving of posts; the cattle are not yet out in the fields, so you can take down the wires; and there is the pleasant sight of apple blossoms to relieve the tedium of your work.

Your neighbor has cleaned out his hoghouse, and spread the contents thereof on a field to windward of you, on the day that your wife is staging a garden party for her city friends. It gives an added topic of conversation.

After fence building, the tractor comes out for plowing of last year's pasture, and fitting it for cornland. Take a little time off to go down to the brook and let the children fish with forked sticks for fresh watercress, eaten in high sandwiches for Sunday night supper. The first straggling rows are up in the garden, and the cows, at last, are out on pasture, acquiring a deeper yellow tint in their increased flow of milk.

The lambs in the orchard have lost their early fleece and long legs, and have almost forgotten how to be playful, they are gaining so much weight. It is time to mix salt with phenothiazine for them, to eliminate stomach worms.

By June the regular garden is furnishing radishes and lettuce, the first greenish-yellow plantings of sweet corn are above ground, and the hayfields are knee-high, with bees busy at the first blossoms. With the tractor wheels set extra-wide to span the rows, corn cultivation is the order of the day. A harness of cultivator hoes makes the machine look like a giant insect as it buzzes around the field.

By mid-June the annual school picnic is out of the way, the children are fully absorbed in their vacation routines, and it is time to make hay between the rainspells. From all sides the wind brings the refreshing smell of newly cut alfalfa and clover, and the former blue and green hayfields show as big rectangles of brown stubble when

the hay is removed. Meadowlarks and pheasants call plaintively, trying to locate their old nests. The swimming hole is in full swing, to
take the dust of haymaking from the sweaty backs of young drivers,
and there's strawberry shortcake on the table. Thistles and other
weeds will soon be ripe for a dose of 2,4-D.

JULY-AUGUST

By the first of July the fattest of the lambs is ready for butchering,
to make the traditional peas, new potatoes and spring lamb dinner
for July Fourth. The garden is swinging into full production, and
there is time for only one more cultivation of corn before the stalks
get too high.

For the next month, things will go easier for the farmer, even if
his wife has to slave at picking and putting up berries and vegetables
for next winter. It is vacation time—a week on a lake in the North
Woods, while a neighbor tends the cows and feeds the dog. By the
time the family returns, the oats are ready for the combine, beginning the harvest round which will not end until long after frost.
After the oats are off, the straw is raked in long windrows, ready
for the baler, and the farmer and his son sharpen the mower blades
for the second cutting of hay. The cows have eaten the pasture
down to the roots, and munch gratefully on cornstalks tossed over
the fence from the bed of the hayrack. Their milk production has
dropped with the advent of flies and hot weather. Broilers from the
spring chicken crop are going to market, and the farm wife pleads
with her menfolks to clean out the chicken house, to make ready
for the thriving pullets.

SEPTEMBER-OCTOBER

With the second cutting of hay in the barn, now filled to its peak,
out comes the corn binder and mows down row after row of juicy
cornstalks, to be chopped into bits and blown into the high silo. The
heftiest son, like a treadmill horse, prances round and round in the

filling silo to pack down the silage as it rises. Air pockets would prevent the proper fermentation of the silage, and the cows would not get their winter tipple if the youth failed in his harvest dance.

Just before frost, pumpkins and squashes go to the root cellar, along with tomato vines, the fruit still on them to ripen later. The unused celery plants are put in sand in the root cellar, and the last of the melons and sweet corn appear on the table. Apple and grape picking are in full swing, and the late potatoes are dug and bagged for the winter.

After frost the standing corn turns brown and rustles in the chill autumn breezes. The corn picker can come any day now, to roar around the fields and fill the wagons with mounds of yellow ears, ready to be shoveled into the crib. With the exception of broccoli, late cabbage and leeks, which thrive almost to snowtime, the sere garden plants and vines are raked up and burned, to destroy any insect or bacterial pests they may harbor. Since the middle of September the cows have been on the lush pasture of new hay seeding, now grown well above the oat stubble. Their milk production has jumped back near the spring level. A wagon box full of ear corn stands in the pig pasture, with a shovel to dish it out to pigs almost ready to be loaded for market.

Barn doors and loose boards on the barn are being nailed tight, ready to ward off winter winds, a snow fence is set up to protect the driveway to the main road, and outside pipes are disconnected and drained. The boys sneak out with the beam and a log-chain on Halloween, to upset the school privy.

NOVEMBER-DECEMBER

Mindful of the hard freeze due soon, the farmer digs the last of the carrots and other root crops from the garden by the first of November. Pigs and cows are turned into the fields stripped by the corn picker, to glean what they can. The first light snow ends outside pasturing of the stock, and the farmer begins the daily round of getting down hay and hauling out manure from the barn.

Pheasant hunting season is over, and some of the fattest birds are in the quick freeze locker, for a winter feast.

The boys test glare ice in ditches along the road, and throw rocks on the ice of the ponds. If they are lucky they will have skating by Thanksgiving. The corn binder cuts the rows untouched by the picker, and the farmer gathers the bundles into big shocks set in regular rows throughout the field.

Thanksgiving itself has a meaning on the farm which it can never achieve in the city, for everything on the table, from goose to pumpkin pie, has been raised by the farm family, not bought at the store. The smartest city relatives are those who come out to the farm to taste the riches of nature.

After Thanksgiving, virtually nothing remains to be done in the fields, except the daily chore of hauling out manure from the barn. All that is left are the rows of corn shocks, which will be brought in as soon as the ground is hard, for shredding as fodder and bedding for the cows. By Christmas, beside a tree cut from his own woodlot, the farmer is ready for another feast of plenty, and the end of a year of satisfying work.

ADDITIONAL READING MATERIAL: *Country Notes*, by V. Sackville West; *The Country Home Month by Month*, by Edward Farrington; *Countryman's Companion*, edited by David B. Greenberg; *The Old Farmer's Almanac*.

Pamphlet: (from Supt. of Documents) *Amateur Forecasting from Cloud Formations*, 5¢, Cat. No. A 1.10/a:1820.

PART II
What You Can Do
WITH ANIMALS

-8-

DAIRY AND BEEF CATTLE

DAIRYING

Can you stand being chained to a cow?

Thirty head are all one man can handle.

Producing for city market is most exacting, a condensery is easier, and raising breeding stock is easiest of all.

"Mail order" bovine romances enable you to build a fine herd cheaply and gradually.

Teaching your cow to lead with a rope will save your temper.

The milking cycle, from artificial insemination to full milk pail.

Two fingers teach a calf to drink.

You won't know whether you are winning or losing unless you test for production, and your family's health may suffer unless you also test for disease.

Be sure to "water the milk" through the cow.

Stanchion vs. loose barns.

The common dairy breeds.

BEEF CATTLE

"The riskiest kind of farming."

Follow beef cattle with pigs to make a profit.

Close figuring on costs of grain and supplements is needed.

Dairy and beef cross for good eating.

Common beef breeds.

New types of barns and hay preservation for dairy and beef cattle.

DAIRYING

"THE man who runs a dairy herd is chained to a cow."

Like many another country saying, this one also is only partly true, but it underscores the fact that successful dairying requires more constant attendance than any other branch of farming. Studies have shown that cows give the maximum when they are milked twice a day, about twelve hours apart, by the same people, on a schedule that doesn't vary more than half an hour from day to day. (I am excluding from consideration here the purebreds on advanced registry who are milked three and four times a day.) This means that if you want maximum production, you will be getting up Sundays at the same time as weekdays, and you will be back in the barn Sunday afternoon, and every other day, at the same time, regardless of the attractiveness of any social doings. Of course, if you don't care about squeezing the most milk out of the cow, you will observe more flexibility in your milking schedule—but you'll have a smaller milk check.

For a successful dairy farm you will need at least seventy or eighty acres of good plowland and good pasture, and it must either be near a large city, on a regular milk truck route, or be close enough to a cheese factory, condensery or other milk manufacturing plant so that you can do your own hauling, if there is no regular milk hauler.

Farms vary so much in quality of land and buildings that only the broadest price limits can be indicated. It seems unlikely, with present postwar prices, that a good eighty acres, with buildings and near a market, can be bought for much less than $10,000, nor is such a farm likely to pay out if it costs more than $20,000. A herd of good cows (a poor cow is a drain on your purse, not an income producer), and the milk and feed-raising machinery necessary for the farm, will add about another $15,000 to the investment, for a total of $25,000, more or less.

You may be chained to a cow, but in the last ten years, it has been a golden chain, thanks to high milk prices. Even with lower milk prices the dairyman who raises most of his own feed can still make a good living, and it is perhaps the surest way to make a living with cattle.

First of all, what constitutes an efficient-sized dairy herd, and how many cows can you handle on a farm your size? Each cow needs between three and four acres of land to maintain herself for a year—the larger figure being for the bigger breeds. This includes pasture and plowed land for raising the grain and hay ration (woods and marsh are of little value for cows). If you are carrying other livestock or poultry, you will have to deduct the land they need from that available for cows.

To calculate land and feed needed for various farm operations, farm experts consider one horse or one cow as a "livestock unit." For other farm animals, the equivalent of one livestock unit is five hogs, seven sheep or 100 chickens. Thus if your land is able to maintain twenty cows, you could substitute ten hogs for two of the cows, and still raise enough feed. Young stock are figured at half a unit.

About thirty head of cattle, of which some will be calves and heifers being raised for herd replacements, are considered the maximum-sized herd that can be handled in a farm unit by one man. Anything considerably smaller does not bring in enough income to warrant investment in labor-saving devices like milking machines, coolers and the like. A herd larger than thirty head means long runs in the barn with feed carts, and other chore jobs, and long lanes and drives to and from pastures and fields—carried on by hired help which spends a good part of its time traveling from one place to another without producing. The big two hundred-cow farm can afford a lot of labor-saving machinery, but the margin of profit for the investment is not as great as that of the thirty-head farm, run by one man and possibly a helper.

Even with a small herd, there are several methods of dairying. You can arrange the breeding of your cows on a staggered basis, so that you have a fairly steady flow of milk throughout the year. This is sometimes desirable when producing for the big city markets, where milk dealers require each farmer to stick rather close to a "base" of production. Farmers supplying the big cities, also, must usually observe a considerable number of rules involving location and construction of milkhouse, upkeep and cleanliness of the barn, and frequent herd testing. If you intend to sell on such a market, you must of course check in advance with the city milk inspector about the requirements, and make sure that you are tied up with a dealer who will accept your milk.

An alternative is producing milk for a cheese factory or condensery. Here the sanitary requirements are not apt to be so strict, and usually there is no year-round "base" production required. For instance, if you want a long vacation in summer, or want full time to devote to field work instead of milking chores, you can breed your cows so that all will calve early in the fall, and after about ten months of production, the herd will be dry during the summer months, with no milking required. Some farmers who follow this practice buy a cheap bull nine months before the desired calving time, letting him run with the herd for a month or six weeks, which will insure breeding of all the cows, and then sell the bull. Of course this method produces calves which are necessary to start a fresh flow of milk, but it does not insure quality breeding stock for replacements in the herd. On the contrary, if they don't want to raise scrub animals into milking cows, which is a losing proposition, such farmers can buy replacement heifers elsewhere, or breed a few cows artificially, if artificial insemination is available in the neighborhood.

Still another method of dairying is not to do any milking at all! If you don't like the looks of a milking machine, or the feel of a cow's teats, you can buy young calves, feed them specially prepared rations obtainable from your feed store, breed them when they are between fifteen and eighteen months old, and then sell them to other farmers as replacement stock, just before the heifers are due to "freshen" (that is, have a calf). A neighbor of mine does just that. He gets both heifers and bull calves, castrating the bull calves and raising them for beef. His fields are always full of young stock growing toward maturity—and he keeps just one cow for milk for his family.

Artificial insemination, where it is available, provides the most economical method of building up a good herd of purebreds without laying out money for expensive stock. It is especially suited to the beginning farmer, who can learn gradually how to handle a herd, without having a whole mass of bovines thrust upon him at once, before he has the strength and skill to cope with them. The farmer under such a program buys one or two very good cows of the type he has chosen—but with due regard, also, to what types of semen are available from the artificial insemination station.

Suppose he has picked Guernseys, and wants to acquire one middle-aged cow of good type, but cheap because of age. How can you pick out a cow of a good type? You can't, when you are just beginning. But every neighborhood has at least one farmer or villager who, for a dollar or two fee, will tell you whether the cow you are thinking of buying is good or not. It is money well spent, and I have used such service myself.

Artificial insemination has three advantages: it gives you the service of a better bull than you could probably afford (without expense of maintenance either), permits breeding young heifers before they are big enough to support the weight of a mature, tested herd sire, and avoids the necessity of maintaining a potentially dangerous animal around the farm. (Incidentally, dairy breed bulls, particularly Jersey and Guernsey, are apt to be "meaner" and more vicious than the more placid beef-type bulls.)

All you have to do is note when the cow is in "season," lock her in the stanchion, and telephone the inseminator. A cow's menstrual cycle is usually three weeks, though some are shorter and some longer. When other animals ride her, she is in season, and ready to be bred. For about twelve hours the cow will ride other animals, and be ridden by them, and it is during the latter part of this twelve hours that she is most likely to conceive if bred. She will also conceive (called "settling") during the next twelve hours, but the percentage of success is slightly less. When you call, the inseminator will want to know what time you first noticed the cow in season, so he can gauge his arrival accordingly.

The process of insemination, done with a long glass rod, is painless and pleasurable to the cow, and takes only a few minutes. All you have to furnish is a pail of water for the inseminator to wash off his long rubber glove. Mark the calendar three weeks ahead, and if the cow is in season again, repeat your call to the inseminator. Most artificial insemination services are sold on the basis of three services to one cow if she does not settle the first time (the average is less than two services to settle).

A new service, begun experimentally in some parts of the country, is a pregnancy examination made about forty-five days after a service which seems to have settled. I realized how valuable this service is when a cow of mine, bred artificially and apparently settled after the third service, unexpectedly came in season six months later, just as I was getting ready to expect a calf. I had lost six months of milk production because of it.

Finding out whether a cow is in season usually requires at least one other animal, or there won't be anything to ride. For the one-cow farmer this extra animal doesn't need to be another cow. It can be a heifer or young calf, or a steer. A man with just one cow and a steer he is planning to turn into beef would hold off on the execution, for instance, until he is sure his cow has settled.

If no artificial insemination is available, or the bulls are not the breed you want, you will have to depend on your own or the neighbor's bull. The best bull, of course, is what is called a "proved" herd sire—one whose daughters have demonstrated good milk production and good type—but such animals are not too common. They may also be too heavy for your cow. You can usually pick up a young bull (they are ready for service at a year) or get a bull calf from a good herd, and raise your own. Sometimes young bulls raised by Four-H Club members are available, and are usually of good stock.

Since bulls are difficult to load into a truck or trailer, and hard to lead along the highway, you will probably have to load or lead your cow, if you are going to depend on the neighbor's bull. Don't wait until the last minute to find out if your cow will lead. Start a week early, putting a rope around her neck, continuing the rope in an inside twist around her muzzle. Don't try leading her without the twist around the muzzle, or she'll pull you all over creation, as even a small calf can do. A few turns around the barnyard and out to the road will soon teach the cow to follow the rope in a docile manner. (I once spent four hours taking a cow half a mile, because I had not taken the trouble ahead of time to teach her how to lead.)

Here is a step-by-step description of handling a cow, which can do for one, or a whole herd:

Let's assume that the cow has been bred, and will have a calf in about two months. Since the calf makes most of its growth during the last two months, the cow should be dry, and get plenty of feed, such as good pasture, or green, leafy hay plus a balanced ration of ground corn and oats plus minerals and protein, or silage. The last two weeks before she is due, the udder begins to expand and should be felt daily to see whether it is hard or inflamed. If it is, call the neighbor or the vet. In the few days before the calf is ready to be dropped, the hip bones spread and the tail slants between them, instead of being carried high. If you have a box stall or other suitable maternity ward, now is the time to put her in, at the same time cutting out the grain and most of the silage. Give her plenty of water and hay.

During birth, the forelegs of the calf, with the head between, appear first, but sometimes, if the cow is small and the calf large, she will have difficulty in expelling it. If she makes no progress in an hour, call your neighbor or the vet. They will put a rope on the front legs, pull in time with the labor contractions, and deliver the new arrival. You will be surprised how hard you can pull. If the calf does not show signs of life, see whether there is mucus or skin covering the nostrils, and whack it on the chest to start breathing. Note also whether the afterbirth is expelled, and if it does not appear in a few hours, call the vet.

Sometimes the cow suffers from milk fever a few hours or a few days after dropping the calf. She refuses to get up, and puts her head on the ground. The cause is a sudden draining of calcium from the blood into the milk, and is most apt to occur in high-producing cows. The old cure used to be to inflate the udder by a bicycle pump applied to the teats. Now, however, the vet is called, and he gives the cow a pint of calcium gluconate or other calcium salt through the veins, which puts her on her feet in half an hour, as good as new.

Be sure the cow has hay and water, preferably warm, but give her no grain for the first day, and start with small amounts the next day, increasing gradually. You don't need to bother about milking her the first day, and it is better to bring the cow up to the maximum milk flow in about three weeks.

The second day, milk just enough morning and night to relieve any tightness in the udder. If you don't know how to milk, the directions are simple. Approaching from the right-hand side of the cow—as all proper cows are approached (no one really knows why) —grasp the teat in the palm of the hand, thumb-side up, and compress the top of the teat with the thumb against the beginning of the forefinger. This traps the milk in the lower portion of the teat. Now squeeze out the milk, keeping the pressure at the top, and working the pressure down by a gradual closing of the hand toward the little finger. You'll squirt a few streams up your coat sleeve and against your legs before you gain control, but in a short time you will be able to make the milk "sing" in the pail. Do as the calf does—bump your hand up into the udder every now and then, to stimulate the flow of milk. Don't grasp above the teat into the udder, or you may injure the milk glands. If you have never milked before, an obliging neighbor will probably let you practice on a drying cow, before your own cow freshens.

You can take your choice of letting the calf suck for a few days, or taking it away immediately. If you let it suck, you'll have to teach it to drink from a pail later, or buy one of the commercial pails with an oversized nipple fixed to the bottom. The first three to five days' fluid are colostrum, not milk, and this reddish liquid is nature's laxative for the new calf. Besides being a laxative, colostrum also contains valuable vitamins which protect the calf against disease in its first few days. Don't try to sell the colostrum as milk, and you probably won't care to use it yourself.

After the second day, start milking out the cow completely, by stripping her. This is done by running thumb and forefinger down each teat at the end of the milking, until no more milk is produced. Unless you milk out the cow completely each time, she will soon dry up. After the milk has cleared of colostrum (no more red tint on top of the pail) put the calf in a separate stall, or you won't get much milk.

Don't try to teach the calf to drink until it has been without food for twelve hours, or it won't have interest enough to co-operate. You can use either one of two methods, depending on your temperament and strength. One method, using the small-sized calf pail, half full of milk, or milk and water at body temperature, is to lock the calf in a special calf-sized stanchion, wet your finger in the milk repeatedly, put it in the calf's mouth, and try to lead the calf thus down into the milk. Some farmers favor two fingers, spread in a **V** so that they pass on either side of the jaw, inside the lips. It takes a lot of patience, and a firm hold on the pail, because the calf will inevitably butt it. The other way, good if you are strong, is to lock the calf in the stanchion, set the pail down in front of it, put one finger in the calf's mouth and the other on top its head, and just shove the head down until the muzzle is in the milk. The calf will not learn to drink immediately, but he will certainly teach you how strong he is. Keep it up, and after that pailful has been splashed all over you, try again the next morning or night, when the calf will be a lot more interested in getting nourishment. After he learns to drink, start him with a little less than half a pail of whole or skim milk, increasing it with addition of some water, until he is taking a full (calf) pail by the sixth week. Also, by the end of the first week, start him on hay, and give him plenty of special calf feed, which can be either the kind sold in bags by commercial feed dealers, or a special ration that your local feed dealer will grind and mix.

Some farmers, anxious to save milk, sell their calves for veal within a week, although this doesn't make good veal, and is called a "deacon."

If you are keeping the calf for beef or replacement, you can stop horn growth by applying a chemical obtainable at your feed store to the horn buds. This has to be done when the calf is about ten days old. If you don't stop the horns at this time, you may have a messy and painful dehorning job to do later. Electrical dehorners are also available now.

By the end of the third week, with gradually increased grain and silage rations and steady milking, the cow should be at her maximum milk production. Depending on the size of the cow, she should produce from eight to a dozen or more quarts at a milking. Whether you use a milking machine or milk by hand, do it fast, and you'll get more milk and avoid udder troubles. A minute and a half before milking, wash the udder with a cloth dipped in warm water. This will clean the udder of surface dirt, and also stimulate the milk flow.

Bovine tuberculosis, mastitis and Bang's disease (contagious abortion or brucellosis) are three troubles you should be prepared against. Many states now have a compulsory test for bovine tuberculosis, which any vet can give. If the cow shows infection, she should be sold for beef at once, and no milk used from her. Mastitis (inflammation of the udder) often shows up in stringy, yellowish-tinted milk.

Mastitis is an infection for which no permanent "sure cure" has yet been discovered. It can be caused by a number of bacterial organisms. Penicillin, the sulfa drugs and streptomycin are effective in clearing up most types but they are no guarantee against reinfection. The best "cure" is to keep mastitis from getting a toehold in the herd, by sanitation and correct milking order. Since the bacteria causing mastitis are around all the time, on hands, teat cups, and usually in the cow's udder, they can multiply quickly if the cow's resistance is lowered. Clean floors, plenty of dry bedding, bigger stalls to prevent crowding and disinfection of hands and teat cups between milking each cow are the chief herd management methods to prevent mastitis. The order of milking is also very important.

Milking heifers first, then the clean cows, and finally the infected cows will do a lot to prevent the spread of infection.

Bang's disease, contagious abortion or brucellosis, can also play havoc with a herd. Progressive states are now testing whole areas to eradicate it, and your veterinary has a vaccination for calves four to seven months old which gives some protection. He can also check the cows to see if any are reactors. Since undulant fever can be contracted from cows with Bang's disease, and is a serious and sometimes fatal ailment, your cows should be checked by the vet every few months for infection. The safest way for the countryman using his own milk or that of a neighbor is to pasteurize the milk. Home electric pasteurizers for half a dozen bottles of milk are now available, or the milk can be heated to 165 degrees for five minutes, and then rapidly cooled.

The cow will come in season again from two to three weeks after she has dropped her calf, but you should wait for the second time in season, before breeding her again, in order to give her a little rest.

After the maximum milk production, which generally lasts a month or so, the cow will gradually decline in output. However, if she had her calf in the winter, she will give more milk when she gets out on fresh pasture, and after a hot summer and plague of flies, she will pick up again on fall pasture. It is possible to get the absolute maximum milk production out of a cow by timing her calf for late summer, so that she gets an added spurt on fall pasture, declines slightly in the winter, and then picks up again on spring pasture just before drying up.

While she is being milked, and while the new calf is growing inside her, the cow receives the regular full ration, which depends somewhat on the feed you have available. Plenty of good, green, leafy hay, cut young when it has the most protein, is basic. To this you can add twenty to fifty pounds of silage and five to ten pounds of ground feed (ground corn, oats, rye or wheat, with a protein supplement such as commercial dairy feed or soybean meal) depending on the size of the cow. Ordinarily, about one pound of grain is fed for each four pounds of milk produced.

(A much more extensive discussion of feeding dairy cows and other animals, with tables of feeding rations, can be found in *Feeds and Feeding*, by Frank B. Morrison of Cornell University, now out of print but still available on many library shelves.)

If you are short of grain feed but have good hay, you can get by with just using hay. Studies have shown that you will get only about 80 per cent of the milk production obtained with added graining.

Within two months, but not less than six weeks, before the next calf is due, the cow should be dried up, so that she will get back in good flesh and have a healthy calf. You can dry her up by just letting her go without milking.

Remember, when feeding dairy cattle, that about half of the feed goes merely to maintain the body. The other half is for milk production. When you cut down on the ration, the cow will still use the same amount for maintaining her body, but will give less milk, in proportion to the decrease in feed. This point is important to bear in mind when you bump into the information that cows make a little more efficient use of their ration when it is scant than when it is liberal. The slight gain in efficiency of nutritive use is not worth the possible loss in milk production from underfeeding.

When you figure chore time, feed, barn-space investment, and other factors, it is a fair question whether you can save money by raising your own herd replacements, instead of buying them, but there is another and more important consideration. The man who raises his own replacements knows just what he's got, and he is likely to have better quality than the man who buys breeding stock from someone else. Look at it this way: Your neighbor, or the fellow two miles down the line, is dependent for a livelihood on his cows. Is he likely to sell you the best heifer in his barn, one that he is sure will produce a lot of milk—and good calves—during her life, or is he more likely to sell you something that he would just as soon not be giving barn room to?

You can figure the milking life of your cow anywhere from six on up to as many as fifteen or more years, although over nine years is unusual. Accurate records of milk and butterfat production can guide you on how long to keep her—or whether to keep her at all! Sometimes an udder ailment or an injury to a teat may cut down the cow so that she is milking only three instead of the four quarters. Some cows compensate, and give almost as much with three quarters as with four, and some cows of good type are worth keeping for the sake of future herd replacement calves, even though their milk production is down a little. And some cows are kept for sentimental reasons.

It is hard for the beginning farmer to achieve the dispassionate air of the veteran dairyman, who seems to ignore the gentleness and beauty of a cow, and to think of her only as a machine for producing milk, but after chasing a skittish heifer out of the cornfield, or trying to find a calf that an old cow has hidden in the woods, you will find yourself becoming a little more cold-blooded about Bossy. They can be mighty provoking creatures. When the time comes that her teeth are no longer much good for cropping grass, and she is racked by asthma and creaking joints, it may be a kindness to send her "over the scale," and get what you can for her as canning beef.

Cows are creatures of habit, just as much as their masters. Get them used to seeing you moving around in the barnyard and in front of them while they are in the stanchions, and they won't be wild. You can start the taming process with your own young stock, so that they are used to being handled. Small calves are usually kept in the barn until they are several weeks old, because they are more easily fed and protected from flies. Later they can run in an orchard or small pasture, where they will not be pushed around or injured by the older cattle. After they are a year old, start bringing them into a regular stanchion with the other cows, so that they get used to dairy cow behavior, and will be docile by the time you are ready to start milking them. Some herdsmen who value their milk checks discourage strangers from coming into the barn, especially at milking time, and even if you permit visitors, you are perfectly within your rights in cautioning them not to walk in front of cows in the stanchions. This is one of the quickest ways of lowering milk production, for cows are very sensitive to strangers, especially in front of them. This applies also to any unusual noises or commotion in the barn at milking time. Another good reason for keeping out strangers is that they can bring disease into the barn.

Whatever breed you choose, butterfat testing and good records will show you whether you are running a charity or a profitable enterprise. Most milk is sold now on a butterfat basis, as well as so much per pound. (There are about 46 quarts in 100 pounds of milk.) Generally, the base price is figured at so much per 100 pounds containing 3.5 per cent butterfat. If your milk has more butterfat, you get a premium of a few cents for each tenth of a per cent increase in butterfat, or a deduction if you fall below 3.5. Individual cows, even of the same breed, will vary as much as one or two per cent in butterfat, for the same amount of feed consumed, so it is important to test each cow regularly, to find out which are the boarders and which are paying the profits. The milk handler to whom you deliver milk will check the cans to find the butterfat output of the herd as a whole, which will govern your price, but the county or local dairy herd improvement association will tell you how each individual cow is doing.

The butterfat test, together with the record chart of the pounds of milk produced by each cow night and morning, will show you which cows are worth keeping, and especially, which cows are likely to transmit high butterfat and milk production to their offspring. Good cows are now producing at least 300 pounds of butterfat a year, the better ones are doing 400, and the top ones are producing 500 or better. Milk production for top-quality cows, thanks to improved breeding and testing methods, now runs 14,000 pounds a year and up to over 20,000.

Remembering that milk is mostly water, be sure your cows get plenty of it, for it has a direct effect on milk production. The way to get the most into the cow is through individual drinking cups in the barn, and you can promote thirst, as many farmers do, with individual salt blocks mounted on the stanchion. If the barn is not equipped with individual cups, and you want to put off that expense, you can have a tank inside the barn, or an outside tank. In either case, be sure in winter that the cows have plenty of time to drink. Sometimes one or two timid cows will be driven away from the tank by others, and never get a chance to fill up.

Our grandfathers, in cold climates, had tanks with a built-in wood stove, to keep the water from freezing, and it was a mean job to keep the stove going. Now you can get a floating electric disk which will keep most of the surface free of ice, or you can install a hydrant-type fountain in the cowyard, heated electrically. Another method, in cases where the cows are out a good part of the time, is to have the tank inside the barn, but against an outside wall, with a door hinged from above, which the cows soon learn to push inward for a drink. The warmer you can keep the water in winter, the more the cows will drink. (Running water, if available, is cheaper than water heated electrically to keep it from freezing.)

Depending on what part of the country you are in, the cows will be in the barn, and be fed stored feeds, for from two to seven months of the year. That also means manure. You can get away from the daily chore of hauling out manure if you keep the cows in what is called a "loose barn," but you will have to check local milk sanitation regulations to determine whether this is permissible. In a loose barn the cows spend most of their time in a three-walled shed, and can wander out into the barnyard at will. Twice a day they are brought into a separate "milking parlor," put in stanchions, milked, and usually given their grain ration.

Studies carried on for several years by the University of Wisconsin College of Agriculture indicate that milk production is about the same whether cows are kept in a stanchion barn or in a loose barn. In a loose barn, however, the cows eat more of the cheaper roughage like hay, instead of the expensive grains and concentrates. They also seem to be freer of disease, and live longer. Chore time is also considerably reduced, because there is no daily necessity of warming up the tractor or hitching the team, to haul out manure. The loose barn however, uses about half again as much bedding as the stanchion barn, and the big hauling-out job comes in spring, when you are busy with other work. The barn itself is considerably cheaper to build, if you are starting fresh.

A new development to aid the herdsman who has a stanchion barn is the "cow trainer," an electrical device which rapidly persuades Bossy to evacuate in the gutter, and thus makes frequent changes of bedding unnecessary. Users of cow trainers report much less time needed for grooming the cows and cleaning udders. The device, shaped something like a coathanger, is suspended above the cow's shoulders. When she "humps" forward to evacuate, the trainer gives her a mild electric shock, and quickly teaches her to take one step backward before letting nature take its course. After a couple of weeks the cow is trained, and the current can be turned off, at least for a time.

What breed of dairy cow you select will depend on many factors, such as whether you just want one or two cows for your own use, whether your milk handler wants high butterfat content, the severity of the climate, availability of good breeding stock, and your own preferences. Here are thumbnail sketches of the leading breeds, and additional descriptive material may be obtained from the national breeder associations:

GUERNSEY—One of the two greatest in number of head among dairy cattle throughout the United States. Medium-sized, predominantly light brown mottled with white, these Channel Island natives are sensitive, and noted for their high butterfat production, coupled with good milk output. A characteristic of Guernseys, as well as Jerseys, is the quick separation of the cream (butterfat) from the milk. For the man who wants just one or two cows for his own use, so that he can skim the cream and make his own butter, the Guernsey fills the bill, as does the Jersey. The yellowness carries over into the flesh, and the fat is tinted light yellow, which some find displeasing to eat—and some don't.

HOLSTEIN-FRIESIAN—Biggest of the dairy cows, and ranking with Guernseys in total numbers. The black and white Holstein is the country's greatest quantity milk producer. The udders sometimes become so large and distended that they almost drag the ground. Unlike Jerseys and Guernseys, with Holsteins the fat globules do not readily separate out of the milk, which must be separated by machine, instead of skimmed. With gradual improvement of the breed, the cows have become larger, and some of the better types now require six feet of standing room between stanchion and gutter, instead of the usual five and a half feet, and three and a half feet of width between stall dividers, instead of three feet. The flesh is good eating, and the fat is white. Calves are large when born, and bring good prices as veal.

JERSEY—Like the Guernsey, a high butterfat producer, but gives a smaller quantity of milk than Guernseys. This small animal is a uniform dark brown, verging on black in some instances. Some dairymen object that the teats are so small that they are difficult for hand milking. As with Guernseys, the calves are very small when born, and are almost valueless for veal.

AYRSHIRE—Originating in Scotland, with a color varying in different animals from white with brown or red patches to red or brown and white, the animal is of medium size, and runs slightly above Holsteins in butterfat percentage.

MILKING SHORTHORN, RED POLLED and BROWN SWISS—The Milking Shorthorn and Red Polled are "dual-purpose" animals, being pretty fair milk producers with the build of a beef animal. They are popular in corn-producing areas where excess stock can be rapidly turned into quality beef if milk prices are low. The Brown Swiss, slightly smaller than a Milking Shorthorn or Holstein, has some of the chunkiness of a beef animal, but is a good milk producer, with a butterfat output ranking close to the Guernsey.

BEEF CATTLE

Beef cattle raising requires a less exacting time schedule than dairying, but includes plenty of work—of a different sort. It takes many bushels of shelled corn to "finish" a steer to the point where it brings top quality price. With tractor cultivation, corn-picking machines and load elevators, however, the work of raising the same quantity of corn is easier than it used to be.

Like the dairyman, the beef cattle full-time farmer can expect to spend somewhere between $10,000 and $20,000 for a farm of between eighty and one hundred acres, and another $10,000 or more for machinery and livestock. Unlike the dairy farmer, he does not need a location near a market which will take his product daily. He can be some distance away from markets, which will make his land cheaper, but somewhat increase the transportation cost of moving his cattle. Like the dairyman, he needs a farm with good pasture and good cropland.

Many beef cattle farmers buy young stock from the Western ranges, and "winter" them in the barnyard feed lot, which is mostly a process of shoving in corn, hay and silage. Usually the young steers, weighing from 400 to 700 pounds, are bought early in the spring, and are put out on good pasture, where they will add from 200 to 300 pounds by the time winter cuts the pasture. Then they are brought into the feed lot, a place near water and protected from the wind, where they are stuffed with corn and hay—and kept from running off the fat.

Beef cattle raising demands considerable skill in judging the market. For this reason, some experts call it one of the most risky types of farming. It takes nearly two years to raise a beef animal from calfhood to the time it is ready for butchering, so the long-term trends are important. After the steer reaches his best weight (between 800 to 1,200 pounds, depending on the breed) he will keep on eating just as much, but he won't gain more weight. Hence it is important to study the local markets, and determine at what season of the year the highest prices are paid. Fall and early winter are usually the lowest times, because of the influx of cattle from the ranges.

Up to twenty or thirty years ago, the standard method of beef cattle raising was to leave them on range, on scant rations, until about two years old or more, and then bring them from the Western ranges to the Cornbelt, where grain was cheaper, and fatten them for market. The modern method is to start fattening the beeves while they are growing, and to market them from a year to eighteen months of age as baby beef or fat yearlings. One reason is that it is easier to keep an animal fat than it is to fatten a lean one. Two other compelling reasons are that consumers prefer smaller cuts of meat with less fat waste, and, perhaps most important of all, the early gains are the most profitable ones. The feed cost for fattening a two-year-old steer is about one-third more, per pound of gain, than in fattening a calf.

The hard fact of beef cattle raising is that the cost of the feed is greater than the selling price of the increased poundage of meat. Then where do the profits come in? There are three factors which help to spell a profit for the farmer with the sharpest pencil. One is the "necessary margin." This is the difference between the purchase price of the young steers and their selling price as fat stock for slaughter. The stockman must get a few cents more per pound when he sells the animals than when he buys them. Often this is around two and a half cents per pound, but it depends on the age of the cattle, and the varying prices of grain, hay, silage, and concentrates such as soybean meal. For instance, the stockman knows that for each 100 pounds of gain with calves, he will need about 470 pounds of grain, for yearlings nearly 600, and for two-year-olds about 675 pounds of grain. By calculating the cost of the grain, and other items like hay and concentrates, he can figure out how much he will have to sell the animals for to break even. This spread, or "necessary margin," covers the cost of feed, labor, and overhead, but it still doesn't necessarily mean a profit.

Pigs following the fattening cattle are the second factor in raising beef cattle for profit. When the steers are fed ear or shelled corn, the pigs will recover something like 5 per cent of the corn, by picking it out of the droppings. A pig weighing from 50 to 150 pounds will pick up the equivalent of 300 pounds of corn in four months of following the beef cattle. You can use one pig to each three calves, and one to each two steers on full corn feed if the corn is shelled, and twice as many with ear corn. The pigs should be replaced with others as soon as they reach market weight, and must also be fed tankage to make most efficient use of the corn.

The third factor is the manure value of the fattening cattle, which takes the place of commercial fertilizers. This manurial value varies with the different kinds of feed, averaging something over three dollars per ton of manure. Extensive tables are available in Morrison's *Feed and Feeding*, showing the manurial value of different feeds and concentrates, which the stockman takes into consideration when figuring the net cost of the concentrates.

It is possible to fatten cattle almost exclusively on roughages like alfalfa and corn silage, with a limited amount of grain and concentrates, but the animals will not bring the top market price, and there is virtually no pork profit. Experts do not recommend the procedure, unless grains are extraordinarily high in price.

There is no one best ration for fattening steers, because of differing ages of the animals, and changing costs of the feeds. On what is called full-fed corn ration, for instance, a two-year-old steer might be given 14 pounds of shelled corn per day, 3 pounds of concentrates, 3 pounds of hay, and about 25 pounds of corn silage, which he would be expected to convert into a little over 2 pounds of meat per day. This, however, is merely a sample illustration, and the reader should consult his county agent, the feed store, or Morrison for a ration suited to local prices and conditions.

If you don't like the work of raising corn, or your land is too rolling for row crops, you can do pretty well fattening steers just on grass pasture. They won't have the "finish" of a corn-fed steer, but you won't have the work or the soil drain of corn. One possible method is to buy small steers in the spring, pasture them all summer, and then sell them in the fall to farmers for feed-lot fattening.

You can even combine some beef cattle raising with dairying. Many farmers, tired of gnawing on "old cow," killed at an advanced age because she wasn't producing enough, now run one or two steers with the dairy herd—and eat like kings.

Another possibility, if you have a Holstein dairy herd, is to keep an Angus bull for breeding everything but replacements. The Angus-Holstein cross produces an all-black hornless cross which has the size and ranginess of the Holstein and some of the chunkiness of the Angus. Marketed at eighteen to twenty months, they can usually be sold at the same price as beef animals. The cross is a little startling to look at, but it makes good meat. An Angus-Jersey cross is also good.

An advantage of the beef breeds in northern climates is that they can do a lot of foraging in winter, which would lower production in a dairy cow. Beef cattle can range a cornfield all winter, gleaning what the picker left—and helping to fertilize the field at the same time. Because much of their feed is converted into heat energy, they do not mind the cold the way dairy cattle do.

The commonest beef breeds are:

HEREFORD—The classical "white face" of the Western plains, very hardy and a little above medium in size. The body is all reddish color, with a white face.

SHORTHORN—Slightly larger than the Hereford, mottled white and brown in color, and sometimes roan.

ABERDEEN-ANGUS—Smallest in size of the beef breeds, all black and hornless. Proponents of the breed insist that it "dresses out" to a higher percentage of meat than any other beef breed.

As with loose barns for dairying, there are also new types of barns for beef cattle raising. The new structures emphasize self-feeding. The haymow runs clear to the floor, and interior walls slant inward near the bottom. There is a walkway between the outside walls and the inner walls, where the beef cattle can promenade, and stick their heads into self-feeders, getting their hay direct from the haymow, without the necessity of forking it into racks for them.

One of the newest developments, suitable for both dairying and beef cattle raising, is a silo-type hay drier. In parts of the country where bringing out the mower seems to bring on the rainclouds, the hay drier will be a boon. Hay is cut in small quantities in the morning, and allowed to wilt a little. Then it is chopped in the early afternoon by a forage harvester, carted in big bins on wheels to the hay drier, and blown by a pipe to the inside of the drier. The drier has a central core or air chamber, through which a fan at the bottom forces air which can also be warmed. The air draft completes the curing of the hay, which remains fresh and green, with most of the leaves still on it.

The great advantage is that hay can be cut and stored the same day, instead of following the normal procedure of cutting and raking one day, then loading a haywagon the next and hauling it into the barn. Under the old system, some of the hay would be too dry by the time it got to the barn, and some too moist, so there was no uniform quality. With a hay drier, the hay is handled entirely by machinery, and is even unloaded into feed carts at the bottom, saving all the work of pitching and mowing away—the heaviest jobs on the farm.

ADDITIONAL READING MATERIAL: For the most comprehensive treatment of feeds and feeding of dairy and beef cattle, see *Feeds and Feeding*, by Frank B. Morrison, chaps. 25 through 29.

Pamphlets: (free from U.S. Dept. of Agriculture) *Care and Management of Dairy Cows*, F 1470; *Dairy Farming for Beginners*, F 1610; *Feeding Cattle for Beef*, F 1549; *Dehorning, Castrating, Branding and Marking Beef Cattle*, F 1600.

-9-

THE THRIFTY PIG

Smartest of farm animals, the pig is also the quickest weight gainer, and converts feed best into human food.

Raise pigs on pasture or on concrete.

Buy weanling pigs, or better yet, raise your own.

A pig can make a better nest than you can, but you can save more piglets than she can, if you are present at farrowing time.

Save prospective brood sows from the lean, worried-looking mothers.

Self-feeders save your work.

The pig wallows in mud because he has few sweat glands with which to keep cool. Fix him up with a wallow, and keep his skin oiled.

Don't feed for fatness, or you won't get the top dollar in the market.

"Titmen" for the part-time farmer.

The final act: load your pigs with a bushel basket.

THE pig has been described, with some justice, as "the smartest animal on the farm, not excepting the owner." With geese, they probably rank as the most astute of domesticated creatures. You can feed a horse too much oats and he will gobble them up and founder himself. A cow who slips her stanchion and gets into the feed bin will eat till she bloats and dies. Chickens will eat moldy grain and succumb.

But the pig is different. His name is a synonym for greed and

63

gluttony, but he is the only animal on the farm who will eat a balanced ration, if all varieties of feed are set before him, and he is also the only animal who will stop eating when he has had enough. He requires less feed, and considerably less total digestible nutrients for each pound of weight gain, than other farm animals. The carcass "dresses out" to a higher percentage, and it is richer in energy than other meats, chiefly because of the high fat content. The pig is also a thrifty animal from the point of view of feeding the human race, because he converts about 20 per cent of the gross energy in the feed he eats into human food, as contrasted with 5 per cent in poultry meat, and only 4 per cent in beef or lambs. He will gain twice as fast, per hundred pounds of live weight, as will fattening calves, and three times as fast as lambs. He doesn't need expensive buildings, produces many young, and reaches market weight quickly.

In the old days of agriculture, the farmer skimmed his milk, took the cream to the butter factory, and fed the skim milk to his hogs from a vile-smelling barrel standing in the hot sun. Skim milk is still a good cheap feed for hogs, if you have it to spare, but it is not the only feed.

The two accepted methods of raising pigs now are on pasture or on concrete, each of which has advantages. Pigs thrive on clover or alfalfa pasture, and will get almost 30 per cent of their food requirements from it. The pasture is supplemented by self-feeders containing grain and an assortment of minerals and proteins. The usual type of feeder has a central hopper, draining into individual feed openings at the sides. Each of the openings has a cover, to keep out rain and prevent wastage, and the pigs very soon learn to lift the lids. When he is on good rich pasture, the pig will select just the right grains and minerals, and in the right quantities, to supplement his pasture diet.

The other method of raising pigs is on a concrete platform. Advantages are better control of diseases, less "runoff" of fat, and no necessity for expensive fencing. The valuable manure can also be easily collected in a pit at the side, keeping the platform clean, and making it easy to load for the fields.

The razorback "hawg" of our forebears, a combination of skin, bones, and fence-breaking ability, has been succeeded by the sleek porker of today, marketed just six months after he is born, running to 35 per cent lard or less, and with hams and bacon tailored to the consumer taste in smaller size.

As with the choice of raising pigs on concrete or pasture, you can start them in two ways also. The quickest, but not necessarily the best, way is to buy weanlings, usually trucked to your door by a pig jockey, and selling for a good round price. They are of doubtful ancestry and nutrition, but it is one way of getting a lot of pigs quickly, and they are already past the high mortality stage.

The second method is to raise pigs from your own brood sows. You can buy sows already bred, or buy gilts (female pigs which have not yet produced their first litter) and buy or borrow a boar to breed them. Many pig raisers like to maintain sows of one breed, and cross them with a boar of another breed, which generally produces a slightly heavier pig. The number of brood sows you can handle depends on the quantity of feed and pasture you have available.

The national average of survival at weaning time is about five pigs per sow. Your chances of making money with pigs depend largely on improving on the national average. On the basis of a gestation period of three months, three weeks and three days, calculate back from the earliest springtime date that is desirable for your neighborhood (six weeks before pasture is available) and put the boar in with the sows at that time. The sow's menstrual cycle is about three weeks, and the boar will breed them when they come in season. The pregnant sow's winter ration should include alfalfa meal, or the pigs are likely to be deformed at birth.

(Pigs bred in late fall are likely to suffer an iron shortage. A few shovels of sod thrown into the pen at farrowing time will correct the deficiency.)

At farrowing time, each sow should have her own pen. If you have a hoghouse, partitions can be easily made. Some breeders prefer the individual houses set out in the pasture. In either case, provide just enough straw to cover the floor an inch or two. When the times comes to farrow, the sow will make her own nest, carrying bits of straw deftly in her mouth, and giving you a dirty look every time you come near.

Plan to be there at farrowing time, even if it means staying up all night. The sow will give you advance warning, when she starts to build her nest, and your presence may save several little pigs. Before she farrows, you will of course get the sow used to seeing you in the pen. Sometimes the sow makes too deep a nest, and the newly-born pigs smother in the excess of straw, or else she rolls on them.

Before farrowing time, see that the pen has a "bumper" along the sides, which can be a two-by-six a foot above the floor and a foot from the wall. This will prevent the sow from squeezing any pigs against the wall. You should also provide an incubator in one corner, by means of a strong diagonal plank a foot from the floor, and with a flap of burlap or canvas at the front. Cover the top, to make an enclosed space, and put in a 40- to 150-watt electric light for extra heat, depending on the weather. Infrared heat lamps can also be used. The little pigs should be placed as soon as born in this incubator, and they will soon run to it naturally to keep warm.

If you are on hand to help at farrowing time, and have provided bumpers and an incubator, you should save considerably more than five pigs per sow. After the first few days the little pigs will pretty well take care of themselves, letting out a terrific squeal if the sow lies on them. During the nursing period the sow should have plenty of feed and water available.

In case you are planning to raise your own brood sows, now is the time to start keeping records. As with any other animal, heredity plays a large part with pigs. The sow which farrows the most pigs, weans the most, and has the most pigs at good weight at marketing time, is the sow to keep. So are her female offspring, for she is likely to transmit these characteristics. Don't be deceived, as many farmers are, by the physical appearance of the sow at weaning time. The sleek, plump sow may be that way because she has raised few pigs and given them little milk at nursing time. On the other hand, the gaunt mother with her ribs showing and a worried look on her snout probably wore herself down taking care of her offspring. The records will tell the story. Look for: high number of pigs in the litter, high number and good weight at weaning time (six weeks), and high number and good weight at marketing time (five and one-half to six months). Good body type in sow and offspring is also a consideration.

If the sows are in good flesh, they can be bred again, immediately after weaning, to produce a fall pig crop.

Weaning time is also about the right time for castration of males which are to be marketed. If you put it off until the animal is near maturity, he will be counted as a "stag," and draw a much lower price. His flesh, whether you sell it or eat it yourself, will be rank and odorous, particularly the bacon.

It is possible, but only just possible, for one person to castrate a pig all by himself, but you are likely to wear the pig and yourself to nervous prostration doing it, because nothing is quite so squirmy and hard to hold as a young pig. The easiest way is to ask your neighbor if he has a good sharp knife, and invite him over to show you how. Very small pigs can be held up by the hind legs, but for larger ones, say those over thirty pounds, it is easier to put them upside down in a trough, sit on them, and hold the rear legs while the other fellow operates. The pig keeps up a continuous squeal from the moment you lay hands on him until he is released. He yells with equal vigor before, during and after the operation, which leads farmers to the comforting belief that the agony is mostly mental.

There are all sorts of prepared commercial feeds on the market for growing pigs. The usual ration consists of ground corn or oats or both, supplemented with a fourth part of "hog balancer," a feed concentrate which contains essential minerals and proteins. You can also make your own concentrate by buying ready-mixed minerals and adding the specified amounts of protein and ground feed. The mineral concentrates and the ground feed should be available in self-feeders, which you can make yourself from plans furnished by the county agricultural agent, or buy ready-made.

If you want to save your pasture from being torn up, or if the fences are not too good, you can ring the pigs, which will confine them strictly to vegetation. A cheap plier-like tool, and ready-curved copper rings, make it a job that one man can do easily. One small ring in the top of the fleshy part of the snout will do the trick, and you should be careful not to get the ring into the cartilage part of the nose.

Pigs have few sweat glands with which to keep cool, which is why you should provide a shady hog wallow for them. They are also apt to become mangy unless you oil their skins. One way of doing this is to buy an "oiler," which consists of a wheel mounted in a bath of oil. The pigs rub against the wheel, it revolves, and oils them. Another way is to soak a burlap bag in old machine oil and tack it around a post in the pigyard, where they can rub against it. Or you can douse them with half a bucket of machine oil while they are all eating at the trough. There are also new chemicals on the market for mange.

In the fall, both big and little pigs can subsist largely on pasture, can "hog down" standing corn, or can glean after the corn picker. The pigs will clean up the 10 per cent of corn—scattered kernels, missed or broken ears, and the like—which the picker did not gather.

Our ancestors sensed the need of vitamins in the pig's diet. If the porkers did not thrive in the cornfield, they drove in a herd of cattle, and let the pigs follow the kine. There is something about the amino acids and other minerals in the manure of cattle which helps the pig to make the best nutritive use of the corn. (See Chap. 8, "Dairy and Beef Cattle," for the "pig profit" in raising steers.)

Watch your drove of pigs daily for early signs of possible disease. If they are listless, unthrifty or "off their feed," call the vet at once, and it may save you several pigs. If hog cholera is at all prevalent in your neighborhood, it would be worth while to inoculate all the young pigs against it. Remembering that a stray dog can bring cholera to your farm just by running across it, you will probably want to shoot at any strays you see. When about fifty pounds weight, pigs should be wormed, either by forcing a capsule down their throats, which involves handling them, or by putting the worming powder in the dry feed, as directed. Be careful, it's poison.

Sometimes young pigs develop an umbilical rupture, which may or may not disappear later. It is unsightly but usually not dangerous, unless it should become strangulated.

Remember in feeding pigs for market that the top price goes to pigs weighing between 200 and 250 pounds, with 220 a very desirable weight. After the pigs get above 250 you suffer a double loss: the rate of gain per pound of feed consumed is less, and the animals are graded down to a lower price per hundred pounds when you sell them.

While the pigs are growing, watch out for "titmen." These are the little fellows who are crowded away from the sow's nipples, and get a slow start. Later they are crowded from the feed troughs, or seem unwilling to eat. Often a titman can be taken into the barn or a small pen, where he will be away from the other pigs, and fed a little whole or skim milk, plus kitchen scraps, until he catches up with the other pigs. If the titman still does not gain, the most profitable thing to do is knock him on the head and bury him, for he will never amount to anything.

Even the small farmer can usually handle one or two pigs with profit. He can buy titmen for a dollar or two from his neighbors, or weanlings from the pig jockey, and raise them on table scraps, skim milk, cleanings from the cow mangers and chicken hoppers, and what roadside clover the children cut. A bag of hog balancer will start him on the road to gaining weight; he survives cold well if he has a dry, draftless house, with enough straw for a bed; and under such conditions he can be raised to greater weight without loss to the owner.

The final act with pigs, which has probably provoked more profanity on the farm than anything else, including balky tractors, is the loading of the pigs for market—or even the loading of one pig into a trailer to go to the butcher for the family. It seems unbelievable that an indolent, comfort-loving pig can be so suddenly transformed into a slippery, energetic bundle of contrary impulses, quick to force a six-inch crack into an escape hatch, violently unwilling to go up the ramp into the tumbril, and full of ear-piercing shrieks. Some farmers, anxious to get the top dollar for pigs in perfect condition, leave the truck in the pigyard for two or three days, make a ramp to the platform, and feed the pigs on the truck, to get them used to it. Others, more violent and sanguine of temper, crowd the pigs into a narrow enclosure, and load them, squealing to high heaven, by grabbing an ear and hind leg. This method is the hardest on both the pig and the owner.

Since a pig of marketable size cannot be driven where he does not want to go, some farmers use the old bushel basket trick. They put a bushel basket over the pig's head. He backs away, and they back him right up into the truck. This is possibly the easiest way of loading pigs.

Many commercial truckers now carry electric "shock sticks," powered by a couple of batteries, which help a lot in loading pigs, and are more effective than the old canvas flails formerly employed to swat the pig into compliance.

If the pen is located where you cannot get near it with a truck or trailer, and you must drive the pig some distance, use a rope. First throw the pig on his side, and then fasten the rope with a slip knot to one of his front feet, near the hoof. The rope will enable you to control the pig by bringing him to earth if he starts to run in the wrong direction.

August and September in the fall, and from March to spring are the most profitable times for marketing pigs. A great rush of pigs to market from October to January, and again in late spring, depresses prices.

Several agricultural colleges are now conducting long-range experiments on breeding new types of pigs which will have more of the desirable qualities sought by packers. It is the same process as is used to produce hybrid corn: working two lines, each with certain desirable traits, through several generations, and then crossing the two final generations, to produce a heavier and more satisfactory product than either of the two immediate ancestors. Since these pigs are hybrids, you would have to obtain and renew breeding stock from professional breeders.

The new hybrid pigs are among the marvels of modern agriculture, but for many years farmers have obtained a somewhat similar effect by simply crossing almost any of the well-known breeds of pigs. Thus a man with good Duroc sows might borrow a Hampshire or Chester White boar from a neighbor, and produce somewhat more thrifty animals than by straight breeding. Of course the progeny would all go to market, and would not be suitable for breeding stock replacement.

Here are the chief breeds of pigs, with brief descriptions of each, the three most popular breeds being listed first:

POLAND CHINA—Spotted black—somewhat inclined to be leggy.

CHESTER WHITE—About the same conformation as Poland China, but white in color.

DUROC—Red in color.

HAMPSHIRE—Black, with a white belt over the shoulders, and a long snout.

BERKSHIRE—Black, with turned-up, short snout, finishes best at not more than 200 pounds.

HEREFORD—Red, with white on head.

TAMWORTH and YORKSHIRE—Long-sided bacon-type hogs.

LANDRACE—Danish pigs, used for line and cross breeding.

ADDITIONAL READING MATERIAL: (free from U.S. Dept. of Agriculture) *Swine Production,* F 1437; *Pork on the Farm, Killing, Curing and Canning,* F 1186. Also Morrison, *Feeds and Feeding,* chaps. 34 and 35.

-IO-

SHEEP CAN BE PROFITABLE

Don't try sheep unless you have good fencing or are willing to build it.

Shelter: three sides and a roof will do.

For convenience and economy, choose a breed common in the neighborhood.

Fall is the best time to start your flock, and is the only time most sheep will breed.

Lambing time requires full attention, but even "dead" lambs can be revived.

Rubber bands are the easiest method of docking and castration.

Alternate several pastures for feeding. Use phenothiazine powder in salt to control stomach parasites.

Let an expert shear the old sheep.

When and where to market.

Goats for milk

SOME time or other every man bumps into a Westerner, or at least somebody who has talked to a Westerner, and gets a load of unkind remarks about sheepherders, most of them scarcely printable. But raising sheep and being a sheepherder are two different things, so we can ignore the prejudices and antipathies of the cattle-grazing fraternity. The fact is that sheep can be a profitable addition to almost any farm.

Before you think seriously about sheep, take a look at your fences

—and at the dogs of the neighborhood. Even with plenty of pasture, sheep have a genius for discovering fence holes, and if one goes through, they'll all follow or break their necks trying. With the time spent in hunting strayed sheep, the danger of their killing themselves by bloating on wet alfalfa or other crops, and the nuisance to your neighbors, there is no use attempting sheep unless you have good fences or are willing to build them.

Sheep can be kept in by a woven wire fence at least four feet high, with two strands of barbed wire at the top, but it is even better to have five feet of woven wire and two strands of barb, making a fence six feet high. The woven wire should be within a couple inches of the ground, and drawn tight, or they will infallibly push out underneath.

Sheep will seldom clear a fence over four and a half feet high, but dogs will, and that is why your sheep fence should be at least five and a half or six feet high. One marauding dog can kill or maim a score of sheep in a night. If you are at the edge of a city or village where there are many dogs, a high fence is absolutely essential. Farther out in the country, if you are away from the main-traveled roads, you may be able to get by with a fence only four and a half feet high.

On the other hand, if you have an old orchard with medium-good fencing, and have not invested in expensive purebreds, your chances of raising sheep with profit are pretty good.

You can even raise sheep without any fences at all—if you want to go to the trouble of making collars for them and staking them out each day. Many a subsistence farmer does just that, keeping two or three sheep to eat the grass along the roadside or on uncultivated land near the house.

No animal on the farm needs less shelter than a sheep. They need a roof to protect them from heavy rain and snow, and three walls to ward off the wind, and that is all. You can use an old shed, part of the barn, or make a shelter with posts and a straw roof, weighted down with heavy branches. It is best, but not essential, to have the opening to the south.

Don't try to keep the sheep warm in the winter. They have a better fur coat than anything you could buy for yourself, and the wool next spring will be heavier and of better quality if their shelter is cold—but not drafty—all winter.

The type of flocks current in the neighborhood should be one strong factor in choosing a breed. If your flock is small, you may want to rent or borrow a ram in the fall, and if you have a breed similar to that of several neighbors, the problem is soon solved. Some of the less popular breeds may seem attractive to you, but remember that you will probably be cutting yourself off from possible sales of surplus lambs and ewes to your neighbors, and you may not get as much for them if you sell for slaughter.

Buyers for packing houses assert that Southdowns and Oxfords give the most meat per pound of live animal. They are easier to fence in than some of the larger breeds. On the other hand, the rangier type, such as Suffolks, give less trouble at lambing time, and can fight off dogs better.

Following are the most common breeds of sheep:

Southdown—A short-legged, rather small sheep, with wool completely covering the face and legs. It produces the small cuts now in popular demand.

Suffolk—A large sheep, producing good wool and plenty of meat. The short hairs of the face and lower legs are black, as are the ears.

Oxford—Considerably heavier than the Southdown, with wool up to the eyes, and a brown muzzle.

Hampshire—Medium heavy, with black nose and ears, and black markings near the eyes.

Shropshire—Medium heavy, and a good wool and meat producer.

Cheviot—Bare-faced past the ears, medium heavy, with uniformly white wool.

Merino and Rambouillet—Raised for their long wool, rather than for meat.

Besides the recognized breeds, of course, there are numerous crossbreds and mongrels of no recognized breed, which are worth considering if the price is right, and you just want a few sheep to eat the spare grass, and for your locker.

Fall is probably the best time to start with sheep, and the easiest way is to begin with bred ewes. The gestation period of a sheep is five months, and you should plan to have your lambs arrive at least a couple of weeks before the grass appears, so as to avoid feeding of grain longer than necessary.

Sheltered racks from which the sheep can eat hay are all you need at first, plus a water trough, such as a vinegar barrel sawed in half and partly sunk into the ground. About a month before the lambs are due to arrive, start giving the ewes about two pounds each per day of oats, placed in a trough.

With the exception of the Dorset breed, which can be mated at any time of the year, all other breeds of sheep mate only in the fall. The ewes come in season, when they will accept the ram, after the nights turn quite cool. It is also a good idea, when the weather turns sharp, to put the ewes out on fresh pasture. This is known as "flushing" the ewes, and stimulates the mating process. Some farmers believe flushing also helps to produce more twins.

When you have picked a date in spring for the lambs to arrive, count back five months, and then put the ram in with the ewes at that time. But before putting in the ram, examine the ewes, to see that there are no tags or excess growth of wool at the tail which would interfere with mating.

If the ram has a heavy growth of wool on the testicles, clip off this excess wool, or all his efforts may be in vain. (The testicles require a temperature several degrees cooler than that of the body, or the semen will be sterile.)

Ewes have a three-week cycle for breeding, so to make sure that all your ewes are bred, the ram should run with them a full month. If you want to get an approximate lambing date, put some lamp black or oil paint on the belly of the ram, and note which ewes are marked in the morning. You can get a rough estimate by putting the paint in the center the first week, and then on the left and right sides for the next two weeks.

Start feeding oats about a month before the first lambs are due. The only time you need to worry about your sheep is at lambing time, but then they will need all your attention. If you have a fully enclosed barn or shed where you can pen up the ewes about to lamb, it will save most of your trouble. A ewe that is about to lamb starts "making bag" (expanding the udders) a few days before lambing, and a day or two beforehand the vulva begins to dilate.

If the ewes are penned up in a dry place, they will pretty much take care of themselves at lambing time. Watch, however, to see that they do not get into difficulties in delivering the lamb. Sometimes a small ewe will have trouble in expelling the lamb. If part of the lamb appears, pull on it, in time with the ewe's contractions. If the ewe is in labor for several hours and the lamb does not appear, call a veterinary or an experienced neighbor.

After the lamb is dropped, there is a critical two or three hours in which the little stranger is cold, wet and sometimes friendless. If it is the ewe's first lamb, she may be scared stiff when the lamb starts to nurse, and will dash around the pen instead of standing still. You can cure this by holding the ewe, getting the lamb close to one of the nipples, and then scratching him over the base of the tail. This stimulates him to nurse.

Sometimes, if a lamb is born outside in the cold and wet, or on a particularly cold night inside, he will appear to be almost dead. You can usually revive him by dipping him into a pail of warm water, being sure to keep his nose out, or by giving him a few drops of warm whisky and milk.

After the lamb has been dipped in water, however, he sometimes loses the distinctive smell by which the mother recognizes him, and she will refuse to accept him when you bring him back. Try holding the ewe and letting the lamb suck for awhile. If she still refuses to accept the lamb, tie her close in a little stall where she cannot move forward, backward or sideways, and let the lamb nurse for a day.

At least one out of four ewes, and sometimes a higher proportion, will produce twins. Occasionally the ewe will refuse to accept one of the twins. If she tosses him out, you can bottle the lamb (warm, diluted cow's milk, six ounces four times a day, with a spoonful of honey in the milk). If the lamb of one of your ewes has died, sometimes you can get the childless mother to accept an unwanted twin, by tying her in a stall and letting the lamb suck, or by skinning the dead lamb and wrapping the fleece around the new lamb, so that the mother recognizes the old smell.

During the first few days of the lambing season, keep an eye on all the ewes and lambs, to make sure that all are still nursing and not rejected. For twins, check to see that the milk supply is adequate, and give the ewe extra oats. Keep up the grain ration at this time, and until the ewes get out on fresh pasture.

Within a week to ten days after they are dropped, the lambs should be docked and the males—unless you are saving them for breeding—should be castrated. If you want the professional tools, a mail order house catalogue or a sheep magazine will tell you where to get them, along with directions for their use.

However, the easiest way to dock sheep's tails is by means of a rubber band. Twist a stout rubber band quite tight over the tail, and about an inch from the base. The tail will shrivel and drop off in about two weeks. Another method is to use a pair of pruning shears.

There are several methods of castration. Westerners on the big ranches use a heated pair of tongs to pinch off the scrotum, which cauterizes the wound at the same time. Another method is to cut off the lower third of the scrotum with a knife or scissors, and then pull or cut the testicles with an emasculating wire knife. Some college of agriculture pamphlets point out that if the sheepherder is working alone, by far the easiest way is to hold the lamb with both hands, and bite off the seminal cord, which helps to stop bleeding.

If these methods are not attractive, rubber bands can be used, in the same way as on the tail.

A castrated male lamb, called a wether, will grow to a heavier weight, and produce more tender meat, than one which is uncastrated. The hindquarters, which furnish the legs for roasting, will benefit especially, and the lamb will not be graded down to a lower class if he is sent to market.

The lambs will start to eat grass and hay after the first week. If the pasture is not ready (grass should be at least four inches high), they can be fed hay from racks. It is better to have three or four small pastures than one big one. Let the sheep into each pasture a week at a time, and this will give the other pastures time to grow up.

Ordinarily, when sheep and lambs are on good pasture they do not need extra grain rations. Sometimes, however, when the lambs are dropped extra early, and are intended for early spring marketing, it is advisable to feed them in a "creep" until good pasture is available. The creep is an arrangement of slatted boards in a corner of the barn, through which the lambs can pass, but which keeps out the old sheep. Have extra-good quality hay in a low rack in the creep, and a low trough for grain feeding, with a board above it to keep the lambs from jumping on top. Two quarts of ground oats, two of corn, and one of oil meal cake make a good mixture, but many other grain mixtures will do as well. The creep gives the early lambs a good start, but it is not worth while unless the lambs will not be on pasture for more than a month.

The commonest parasite of sheep is the stomach worm, which cuts the weight and vigor of the animal. Larvae of the parasite climb grass stems, and thus get into the sheep's stomach. A mixture of phenothiazine powder in loose salt, left constantly before the sheep, will eliminate the stomach worms. The powder can be obtained at any feed store, and proportions for mixing are stated on the package. The salt feeder should have a roof shelter, or the salt will wash away with the rain.

In late spring, but before the hot weather of June, the sheep one year or older should be sheared. This is a job for an expert. You may be able to find a neighbor with power shears, or he or the county agent or feed store can direct you to a professional shearer. After you have seen it done a few times, and perhaps practiced on a dead sheep, you may want to try your hand at it yourself, but it is a job that requires much skill.

For shearing, the sheep is first thrown to the ground by seizing the wool on one side. It is then upended on the tail, with the head slanting backward a little past the vertical. In this position the sheep is helpless and quiet. You can also use this procedure just before lambing time, to remove any tags of wool near the udder, so that the lamb has nothing but the nipple to hit when he dives under the sheep.

Watch carefully in shearing of ewes to see that neither of the nipples is sheared off. If this happens, the ewe should be marketed.

To get the highest prices, early spring lambs should be marketed at four or five months, but not later than July. After July, the bulk of spring lambs start moving to market, and depress the price. Lambs dropped too late in the spring to market in July should be held until late in the fall, after the market has recovered from the influx of Western lambs.

Feeding out Western lambs during the early fall and winter months is a common Cornbelt practice, and is similar to the feeding of beef cattle. Like cattle fattening, it requires a "necessary margin" or spread between the buying and selling price, and a shrewd job of flock management.

Although milk goats are not important commercially in the United States, they are well suited to furnishing milk to persons on acreages too small to support a cow. They can secure a great deal of their feed from kitchen and garden wastes, lawn clippings, and browsing on rocky slopes. A purebred doe will produce three quarts of milk a day, and about one thousand pounds during the year. A disadvantage is that their lactation period is shorter than for a cow, and most of them freshen in the spring, which means that it is impossible to obtain a uniform flow of milk throughout the year. They require about the same amount of feed per hundred pounds of milk produced as a cow.

ADDITIONAL READING MATERIAL: (free from U.S. Dept. of Agriculture) *Farm Sheep Raising for Beginners*, F 840; (from Supt. of Documents) *Preparing Wool for Market*, 5¢, L 92. Also Morrison, *Feeds and Feeding*, chaps. 30 and 31.

-II-

THE POULTRY YARD—AND SPECIALTIES

The soldier's dream—a chicken ranch.
Housing each bird costs $4.00.
Nothing beats the sheer guts of a hen.
How to tell when a "cluck" means business.
Chicks for the children, but not for the garden.
Pale legs, good layers; yellow legs, boarders.
Thorough sanitation is better than expensive medicines.
"Battery" chickens can't run off fat.
The egg route tests the conscience of the farmer.
Use a hen to raise baby ducklings.
Don't try to mix turkeys and chickens—the turkeys will die.
Hamsters, mink, pheasants and other specialties.

"THIRTY or three hundred" is still a good rule for poultry numbers, and the beginner would be wise to start with the thirty. A survey of soldiers during World War II indicated that the most common postwar dream was to retire to the country and start a chicken farm. How many actually got out to the coops later, and how many lost their shirts in the process has not been determined, but it is a safe bet that a great many failed, because quantity chicken raising is a highly specialized art, requiring much skill and knowledge. Taking care of a thousand birds is a full-time job for one man.

Suppose you want to start at the other end, and learn about chickens while keeping your investment low. Twenty to thirty hens will give you a world of experience, and if managed right, may even show a profit. The first thing to determine is whether you are raising chickens for eggs or meat, because you won't get both. The white Leghorn hen is the greatest egg-producing machine the world has ever seen, but it is not much on meat. On the other hand, the Rocks, both Barred and White, Rhode Island Reds, Hampshires and Jersey Giants run five to eight pounds at maturity, but don't lay quite so many eggs as a Leghorn. While the heavier breeds eat slightly more grain, and lay fewer eggs, they bring more when marketed, either as broilers or mature birds, so you are likely to come out about the same whichever breed you adopt. A prime consideration would be whether you are pointing for an egg market or a chicken market.

If you are thinking of going full-time into the chicken business, you can estimate an investment of about $4.00 per bird for housing them, and you need between 1,000 and 1,200 birds to operate full-time. This would mean an outlay of almost $5,000 for chicken shelters and all that goes with them, plus the cost of the land. Bought from a hatchery, the birds will cost from a dime to fifteen cents apiece as day-old chicks, and each will eat about a dollar's worth of feed or more, before you begin to get any money back. If you figure this chick and feed cost at about $1,500, the total investment before you begin to collect anything runs to about $6,500.

You can raise chickens in a wilderness, where land is cheap, but you will probably more than make up the difference in land cost by extra outlays for transportation, and somewhat lower prices at your nearest market for chickens and eggs. The most profitable chicken ranches are generally located on the outskirts of large cities, where there is a steady market for chickens and eggs. Land costs are somewhat higher, but income is also higher—and steadier. Part of the investment in housing the chickens may be taken care of by a mortgage, and the experienced poultryman can finance his chick-raising activities through the "feeding out" programs available at banks and through feed merchants. (See Chap. 21, *Where to Get the Money*.)

Care of the flock has traditionally been the job of the farmer's wife—and she showed nearly 100 per cent profit, thanks to lack of bookkeeping. Feed came from the oatbin, the corncrib and kitchen scraps, and the "egg money" was usually hers to spend. Sometimes it was the only cash she ever saw. If you don't have a full oatbin handy, it means buying special chicken mash, fortified with all the vitamins and even cod-liver oil, plus quantities of scratch feed, and ground oyster shells.

It is easy to get into the chicken business—and either easy or difficult to get out, depending on how you feel about wrestling with some of the most ornery creatures on the face of the earth. For pure cussedness nothing beats a hen. They would rather steal a nest than lay in the ones you make, they roost in trees instead of in your henhouse, and they'll eat each other rather than the expensive mash you have poured out for them so lavishly. And then, for no reason at all, apparently, they will turn around and lay like mad, keeping it up until you wouldn't think they had another egg left inside them. Because chickens present such a bundle of challenges, some people can't bear to part with them, and others can't get rid of them soon enough. The happiest man with chickens is the one who does his best for them, and then just relaxes.

If you have children, one of the most fascinating ways to start with chickens is by means of one or more setting hens and a clutch of fertilized eggs. Any near-by farmer will furnish you with a "broody" hen—at a price, of course—and usually also has fertilized eggs. The setting hen pecks and says nothing, instead of running and squawking. To determine whether she really has her mind on her business, and is not just dallying with motherhood, stick your fist under her. If she "works," that is, spreads her wings slightly, and moves her legs to cover what she thinks is an egg, she's in earnest.

Depending on the size of the eggs and the size of the bird, a setting hen can cover from a dozen to fifteen eggs. The nest can be an old applebox, with a few inches of straw, or an old bushel basket with a sack in it. If you are moving the hen to strange surroundings, it is a good idea to put a board or lid on the nest, weighted with a stone, just to be sure she doesn't change her mind. After a couple of days, take the lid off night and morning, so that she won't soil her nest. Have water and feed near by. If she doesn't get back on the nest in ten minutes, put her back and close the lid. In twenty-one days, assuming that the eggs were fertile, your children will be crowing over the arrival of fluffy baby chicks.

They will need growing mash and water, and the old hen will do the rest. If you have flower beds or garden which you don't want scratched, put the hen in an A-shaped box with slats across the front. You may lose a few chicks to rats and other accidents, but the survival rate is pretty good. When the chicks are six weeks old, you can eat the old hen—if the children will let you.

Buying baby chicks from the hatchery takes more preparation as a method of starting out with chickens. For their first two weeks they require an available temperature of about 90 degrees to which they can retire when they feel cold. Farmers with large flocks have a regular brooder house, complete with little oil or electric stove and "hover," which is a canopy of sheet metal under which the chicks can cluster. Chick-sized feed hoppers and waterers are placed just outside the canopy. The floor is covered with litter, which can be bought, or which can consist of straw chaff and sweepings from the haymow. Start with the mash hoppers close to or slightly under the hover, moving them farther away as the chicks become accustomed to running out into the cold.

You don't need a regular brooder house and brooder for small numbers of chicks. An oversized box, eight inches high, with litter on top and burlap strips at the edges to keep out drafts, and with a 50-watt light bulb inside for heat, will do for a score or more chicks. Checking the temperature occasionally will help to save chicks. If it is too cold, they will pile up and smother those underneath.

At six weeks they will use perches, and at eight weeks you can put them out in the field in a "range shelter." This is just a framework, covered with wire, and with pitched roof coming to within about eighteen inches of the ground, which can be made out of scrap lumber from plans available through the county agricultural agent. At eight weeks the cockerels have clearly distinguishable combs, are crowing falsetto, and are usually ready to be converted into broilers.

The third method of starting with chickens is to buy anything from birds eight weeks old up to mature fowls, depending on price, time of year, and available shelter. If you are buying mature birds, ready to lay, pick those with pale legs. Those with bright yellow legs are boarders, who may lay sometime, but not until they have eaten great quantities of expensive feed. A further check is to examine the vent. In laying birds it is moist and wide; in nonlayers dry and round. A fourth test is to place the fingers pointing toward the vent, and about an inch short of it. If there is room for three fingers between the last two ribs, the bird is laying. Sometimes, with gentle pressure, you can feel the next egg. Early moulters should also be culled.

These simple tests to determine which members of your flock are laying and which are just passengers should be employed about once a month to cull the flock. To maintain profits, butcher the nonlayers before they eat any more feed.

If you have bought birds which are ready to lay, or if your chicks are getting toward maturity, they will need a regular laying house. The better the house, the easier it is to keep clean and maintain high egg production, but you can get by with surprisingly little. The former ideal chicken shelter was a "strawloft" henhouse, which was nearly eight feet high at the front end, and had a rack a few feet from the floor, to hold straw. The purpose of the straw was to conserve heat and to absorb the moisture given off by chickens during the winter. The house, made of boards, was expensive to build, hard to move, and not easy to keep clean. One of the outstanding benefits of the agricultural revolution has been a new type of chicken house, built in Quonset shape, with good insulation and ample ventilation to carry off the moisture, eliminating the need for a high roof and strawloft. The houses are light, relatively inexpensive, and very easy to keep clean. They can be used as brooder houses, range shelters, or laying houses, and are easily hauled from one location to another.

If you are not on a fluid milk market, and merely have a cow for your own use, or no cow at all, a corner of the barn can be wired in for the layers. You'll need perches (eight inches per bird) above a dropping board, and a nest for each six or eight birds. You can make your own feeders or buy commercial ones. To give the birds exercise and keep down excessive moisture, have about four inches of loose straw on the floor, adding to it as it gets matted down. In this way you need only clean the chicken house once a year—and there is always the hope that you can persuade some city visitor that it would be fun to clean out the chicken house. The material makes wonderful fertilizer.

During the winter, besides laying mash, scratch feed and oyster shells always available in hoppers, you can increase egg production by a hot mash (hot water and laying mash, stirred to a goo) at noon, and by turning lights on early and leaving them on late.

In the fall, old broccoli and cabbage stalks from the garden can be hung on strings just above the birds' heads, to give them exercise and greens. During the winter, hang up a wire basket of alfalfa or clover hay. An occasional cob of corn will give the birds something to do pecking off the kernels, but too many will make them fat and nonproductive.

Be guided by the big commercial poultrymen: they depend on thorough sanitation, instead of medicines, to keep diseases down and production up. This means periodic cleaning of the chicken house, followed by spraying with any of the common poultry germicides. The diseases of chickens are numerous and often deadly, but the most serious you will probably have to worry about is coccidiosis (evidenced by bloody droppings). There are many commercial preparations on the market to control it, most of them of doubtful benefit, but some poultry raisers claim success merely by the daily addition of a couple of tablespoons of vinegar to the drinking water.

If the prospect of so much sanitation appalls you, you can experiment with a new thought current among some poultry raisers which holds to the theory that it is better *not* to change the litter, so long as it does not become wet. This school of thought holds that old litter develops molds and fungus growths which provide the mysterious "protein factor" essential to thrifty animal life. The chickens, scratching and eating in the old litter, eat this protein factor, and are also benefited by various anti-biotics in the litter growths which attack disease. Persons who have carried on the experiments claim that chickens are healthier and produce more eggs than under the rigid sanitation method.

One further method of raising chickens for meat has become popular in recent years, and that is the "battery" way. The battery consists of four tiers of cages, one above the other. Because they live on wire entirely, the chickens are remarkably free from disease, are easy to keep clean, and are very tender because they don't get much exercise. The battery can also be used as a brooder, and should be considered by anyone wishing to raise birds for meat or market with a minimum of care, dirt, and space.

You can take your choice of letting the chickens run wild around the farm, or keeping them penned up. Running free, they will pick up a good 20 per cent of their feed ration in bugs and grass, thus saving on feed costs. They will also ruin your garden unless it is well fenced, and will make dusting holes under your favorite shrubs, scatter stray feathers on your prize lawn, and leave unpleasant tokens all over the front and back stoop. You can avoid these annoyances by having a chicken-tight yard next to the laying house, but then you will have to furnish all the feed, and keep the yard reasonably clean to avoid disease. Farmers who annually take on five hundred or so baby chicks each spring, and keep several hundred in the laying house during the winter, often just give up trying to keep the growing chickens fenced in. They either move the range shelters far from the house (which invites visits from foxes and other predators, plus long carries of feed and water) or they resign themselves to living in a mess, and just fence in the kitchen garden.

If your flock is bigger than is necessary to supply your family needs, you'll have the problem of marketing eggs and birds. You'll get the lowest price if you take ungraded eggs to the village grocery and sell or "trade" them. The grocer is merely a pickup station for the big egg dealers, and he has to make his profit somewhere. He expects a big slice of it to come out of your eggs. A slightly better price can be obtained by taking fresh eggs, candled and graded for size, to a city grocery or large egg wholesaler. However, these outlets usually accept nothing less than a crate (thirty dozen).

The most profitable way, if you are near a good-sized village or city, is to peddle the eggs yourself on an "egg route." You get a premium above the retail price for strictly fresh eggs, and you can take orders, at top prices, for dressed chickens. What you are doing, in effect, is selling your time—the minutes spent in traveling the egg route, and in cleaning and plucking birds. Incidentally, in the face of uncritical city buyers, many farmers have long since given up wrestling with their consciences as to what constitutes a large-, medium- or small-sized egg. The wholesale buyer will weigh the case to determine the egg size, but the housewife either has no scales, doesn't know what the weight should be for the various sizes, or doesn't want to take the trouble. This principle, or lack of it, is the foundation of much of the profits in egg routes.

In addition to poultry, there are many other specialties which can be added to farm life—some to increase the size and variety of the larder; others, after much more careful consideration, intended to show a profit. Many can give an outlet for energies, as well as priceless training, to the children, even though they just break even or show a small loss. If you become outstanding at it, you may even find a side-line specialty developing into a full-time business.

For home food, geese, ducks and rabbits are easy to raise and help to reduce living costs. Of the three, geese are the hardiest and take the least care. All they ask is plenty of grass and a small water hole. A barrel on its side serves as nest, and an open shed for shelter even during a cold winter. A drawback is their noise, for they can hear a car or even a bicycle half a mile away, and give tongue at once.

Perhaps the easiest way to raise ducks is with a setting hen. The brooding period for duck eggs is four weeks, and White or Barred Rocks make the best setters. Leghorns are more flighty and are apt to leave the eggs when nothing happens at the third week. At the end of the third week, moisten the eggs daily, or the ducklings will not have the strength to break the hard shells. (In nature, the eggs are naturally moistened when the duck returns to the nest daily from a brief excursion to the pond or through wet grass.)

I am not particularly fond of duck to eat, but we always have at least one setting of duck eggs, because of the delight of the children in watching the fluffy ducklings, and the sedate antics of the mature birds. They are worth raising if for no other reason.

The ducklings use the hen as a brooder, and will eat vast quantities of stale bread, skim milk and other leavings of the table. Unlike chickens, they want their food wet, and like to puddle their food with their bills in a near-by pan of water. Until they get their feathers and develop oil glands, the ducklings need no other water. After feathering they need a small water hole. They are even messier than chickens unless penned up, but a much lower fence will keep them. As they near maturity they do well on grass and bugs, but if raised commercially, will do better on special feeds.

If you are considering raising turkeys commercially, or if you just want a few birds for the holidays or to put in the deep freeze, the best way to start is by buying young poults and raising them entirely on wire. This is the foolproof method developed following the discovery that the germs of blackhead, a devastating turkey disease, are carried in a worm parasite of chickens, and appear in the droppings. At present the only successful method of raising turkeys is to keep them completely divorced from chickens. Keep them off ground where chickens have been, by raising them on wire porches. It is also possible to raise them on the ground, but it must be soil where chickens have not roamed for at least a year, nor where chicken manure has been spread. If turkeys are raised on such noninfected ground, they must be enclosed with a five-foot fence, or they will soon be up around the farmhouse. In other words, most of them will soon be dead of blackhead.

When running on open ground, the turkeys are moved about every two weeks to new ground. They do well with a shelter open on three sides, even in winter. Many turkey raisers favored with a hillside start their young poults at the bottom of the hill, and move them up toward the top at two-week intervals, so that the wash of rain over the soiled ground is away from the turkeys, instead of toward them.

The old theory of turkey farming used to be that the birds did not need any special feeding after the first few weeks. They were supposed to forage for themselves. Theories were also held that if the young turks got wet feet, or ate corn in the milk stage, they were likely to die. Successful turkey raisers using modern scientific methods have disproved all three theories. It is now recognized that turkeys make their best growth on mash and scratch feed. Getting their feet wet may lower their resistance somewhat, but is not in itself a cause of death. The soft corn theory has also been shown to be coincidence rather than fact. The big feed companies have developed special turkey mashes which are entirely satisfactory.

There is no particular "best" breed of turkey, but the broad-breasted bronze has been gaining recently in popularity. Breeders are also striving for a slightly smaller turkey, to match consumer preferences.

Some of the other specialties, which require sale to others, need a lot of consideration before you go into them. The fur-bearing animals, such as fox and mink, call for expensive caging and much technical knowledge. With a luxury clientele, and fast-changing tastes, you may find yourself with a big output at a time of rock-bottom prices, or you may guess wrong on strains to breed. Sales of breeding stock are more profitable than raising furs—if the particular strain you are specializing in is in wide demand.

Hobbies like raising bantams, pheasants, hamsters and the like, or breeding dogs, depend largely on being on a well-traveled road near a large city, or on specially developed local markets. Incidentally, if you have the time and the wire for runways, boarding dogs for city owners can bring in a steady income, though it does not necessarily make you popular with near neighbors.

ADDITIONAL READING MATERIAL: *The Care of Rabbits*, N.Y. Agricultural College, Cornell, Utica, N.Y.; *Talking Turkey*, by W. A. Billings, University of Minnesota College of Agriculture, St. Paul, Minn.; also Morrison, *Feeds and Feeding*, chaps. 36 and 37.

What You Can Do

WITH THE LAND

WHAT THE REVOLUTION IN AGRICULTURE ACCOMPLISHED

The revolution helped the war with a "miracle."
Even the wheelbarrow is different these days.
The tractor ends "lost harvests."
Machines which beat the weather and the clock.
Making hay from a tractor seat, with baler or forage harvester.
New barns hold more, lighten the job.
The animals look different too—hybrids, bigger breeds.
Seeds that double the former yields. Corn moves up from king to emperor.
"Fast milking" and a hydraulic lift.
An electric trainer teaches Bossy manners.
Cow romances by air mail.
But there's still a lag in thinking on soil conservation.

ONE could make out a good case that the revolution in agriculture really came of age during World War II. While shipyards bulged with extra help and war plants put on men and women by the thousands, agriculture saw a double drain of youths to the armed services and to the cities. Yet the farms of America, with some 10 per cent fewer persons engaged in agriculture, actually turned out more than during the preceding peacetime years. Part of it, to be sure, represented patriotism and hard work. By the end of the war, the average age of those engaged in agriculture was ten years older than before the war. The old folks just stayed at work, to feed their boys in uniform overseas, to hold the old farmplace until they returned, and to pay off the mortgage with the high wartime prices.

The best of human efforts, however, would not have produced the 15 per cent increase in production—the unbelievable miracle of feeding a wartime population better than in peacetime, while at the same time exporting great quantities of food overseas—had it not been for the quiet revolution in agriculture which had been going on in the preceding years. It was a revolution on many fronts, and it is still going on.

Perhaps the most obvious is the revolution in machinery, and here rubber is king. Possibly the most dramatic example of the revolution is the wheelbarrow. The old iron-wheeled barrow of our fathers, rather small because a big load could not be pushed on a narrow iron tire, has been succeeded by a jumbo of doubled capacity with a pneumatic tire, which carries twice the load with less than half the effort. The iron-wheeled hay wagon and rack, high above the ground, with consequent difficulty of pitching and loading, has a modern brother on four rubber wheels, which hugs the ground, turns in its own length, and can travel over the roughest field.

The rubber tractor tire, permitting high-speed hauling of loads to and from the fields, and along the highways, plus cultivation without packing the soil, is taken for granted now, but it made it possible for one man to cultivate eight times as much corn in a day as a man and team could do before. Rubber seals the equipment in the dairy milkhouse, adds wings to milking machines, and is indispensable in the machines that have taken much of the drudgery from farm work, while they have increased the capacity of one man to do much more in field and barn.

Some of the machines are designed to beat the weather or clock, while others take the load off the farmer's back. In the days of horses a farmer might miss the three or four favorable days in spring when he could put in his oats, and as a result would get half a crop or none. Now, with the speed of a tractor, he can always be sure of getting them in, for a day and a half will do it. New haying equipment beats both clock and weather. Haying, once the hardest job on the farm, can now be done entirely from a tractor seat. The mower, either pulled or attached directly to the tractor, can "knock down" in a couple of early morning hours all the hay that can be put up in a day. Side-wheel rakes, making a continuous swath around the field, replace the old dump rake. After the hay is in windrows, two new types of machine can handle it. Some farmers now use a hay crusher mower which breaks the stalks so that they dry out as fast as the leaves.

Another great new invention is the forage harvester. Attached to a tractor, it marches along the windrow, chops up the hay into lengths of a few inches, and blows them into a big hopper-type wagon pulled behind. The wagon, with slanting floor and door near the bottom, empties with a minimum of shoveling into a blower which wafts the chopped hay into the barn, where it takes up only half the space of the old conventional loose hay. (Many of the new wagons are self-unloading.) In other words, with a forage harvester, you can get twice as much hay in the same-size barn as you can loading it loose.

Sometimes you need to check the beams and flooring, before trying chopped hay. A neighbor of mine, the first in this area to purchase a forage harvester, stuffed his ancient barn with so much chopped hay that the haymow floor collapsed—fortunately when the cows were not in the barn.

Corn pickers that zip over the highway from farm to farm, trailing three picker wagons; "Cornbines" which pick the corn and shred the stalks; one-man combines to harvest oats, other grains, and clover, alfalfa or soybean seed; corn planters with fertilizer attachment, quack diggers, cultipackers (corrugated rollers) to firm the soil after planting so no moisture is lost, one-man string hay or wire balers, portable elevators—these are among the new armament already in use to ease the burden of farm work. One of the greatest of the new back-savers is the hydraulic tractor lift, a pronged scoop on the front of the tractor which can stack hay, load manure, move rocks and dirt, and perform dozens of other heavy farm chores.

On the horizon, nearing perfection, are things like mow-curing of hay, to checkmate the weather; hay driers in the shape of silos, which permit the farmer to cut and store his hay between sunup and sundown, with virtually no danger of rain spoilage; new types of forced draught cribs for hybrid corn to remove the excess moisture which is a characteristic of the hybrid.

The war, with its popularization of the Quonset type of building, also produced new methods of storing machinery. The old type of machine shed, never built large enough to hold all the farmer's equipment, which required a lot of pulling and hauling to nest things in or get them out, has been succeeded by the Quonset-type machine shed, where the farmer drives right through from end to end on the high side, with space off the driveway for all his machinery. The same building converts quickly to a storehouse for surplus hay or grain, and can be used, in a pinch, for livestock. Such flexibility is doing wonders in easing the time-consuming aspects of farming.

The new barns are different, too, from the kind grandfather used. Laminated-wood rafters spring from the foundation in a bulging curve to the peak, permitting greater storage and simpler construction than the old vertical-wall, slant-roof barn. Instead of heavy crossbeams and braces breaking up the interior, the new construction is completely free of barriers, from wall to wall and from floor to roof.

Some barns are being made now on a self-feeding principle. A walkway for stock, just inside the outer walls, opens onto mangers cut into the haymow, where gravity keeps the mangers filled. With beef cattle and an automatic watering system, a farmer could practically spend the winter in Florida!

In some parts of the country, "loose barns" are replacing the conventional stanchion-type barn for dairying. The cows, in a three-sided enclosure, are brought into an easily cleaned milking parlor for milking and graining. Twice a year, with the hydraulic lift on his tractor, the farmer cleans out the accumulated manure in the loose barn in jig time.

Even the humble poultry house has been modernized. Instead of the cumbersome "strawloft" type, made extra heavy and high to provide a blanket of moisture-absorbing straw above, the new poultry houses are low, well insulated, easily convertible from brooder house to broiler to laying hens, and can be hauled from one location to another without any trouble.

New types of animals, also, are going into these new buildings. Hybrid hogs, developed from selected strains to produce a quick-fattening animal of the type the packers want, are a project of several universities, in co-operation with progressive farmers. The U.S. Department of Agriculture's big experiment station in Maryland has been working on hybrid and crossbred cattle to produce more milk and meat. Brahma bulls are widely used in the South for beef breeds.

Hand in hand with hybridizing and crossbreeding has gone much research and breeding of better types of the old established cattle lines. Instead of the national average of 3,000 pounds of milk per cow per year, the average now is close to 6,000. The best animals have zoomed to 15,000 to 20,000 pounds or more of milk per year. Instead of being satisfied with 250 pounds of butterfat per year, the modern dairy farmer expects 350, and many have progressed into the 400- and even 500-pound class.

Chickens also are often hybrids or crossbred, to obtain the vigor and quick-growing qualities produced by such matings, plus the heavier birds which result.

The feed for animals and poultry is different too. The agricultural experiment stations, run under federal-state supervision in connection with the land grant agricultural colleges, are turning out new strains of oats with stronger stalks, heavier heads, and highly resistant to the diseases prevalent in their localities. Do you want plenty of oats for feed? Your nearest experiment station has a list of certified seed growers who will furnish you with heavy-yielding varieties. Perhaps you want more straw for heavy bedding of dairy cattle.

They have that kind too. Even with some of the plant's energy going into longer stalks and hence more straw, yields are double or more above the former national average.

Other small grains have gone through similar developments. Rye and wheat yield more, and are highly resistant to the ailments which used to decimate the crops.

The most outstanding revolution has been accomplished in corn. The old open-pollinated varieties, with uneven stand, odd-sized, small ears, and fragile stalk, have been almost universally succeeded by the modern hybrid wonders. Virtually every stalk reaches the same good height, and it is the boast of hybrid seed corn dealers that you could shoot a rifle down any row and detach every ear, they are hung so evenly above the ground. There is a pardonable exaggeration in this boast, but anyone who compares the ragged appearance of ears and stalks in an open-pollinated field with a hybrid stand would be almost ready to believe it.

Hybrid seed corn is dried artificially, sized to drop evenly through the planter holes, and is virtually uniform in size of ears, number of kernels, and height of stalk. The business of producing this seed corn through generations of matings of desired strains is now one of the major farm enterprises of the Middle West. Corn is also timed as to growing season, for the area in which it is to be planted. Special strains mature in ninety days or less for the early-frost northern regions. Farmers near the Canadian border who took a very sporting chance of harvest in planting any corn at all, now have a fair certainty of a crop, without depending on the less satisfactory "flint" corn. Like the isometric bars of a weather map, a hybrid seed corn chart of the United States shows curving lines, denoting areas where 85-day, 90-, 95-, 100-, 110- and 115-day hybrid corn can be planted with reasonable certainty of a mature harvest.

As with corn and small grains, many another farm and garden crop has benefited from the impact of constant experimentation. Hay used to be timothy and red clover, with an occasional barn filled with oats cut green if the clover winter-killed. Now strains of alfalfa are available which can stand severe winters. Agricultural

experiment stations have discovered that a combination of alfalfa
and brome grass outyields alfalfa raised alone, and makes a highly
palatable mixture. The brome grass fills in the spaces between the
alfalfa roots, and the stand is good for several years, instead of
dying out at the end of the second year, as clover does. Sudan
grass is now a common midsummer pasture crop which can be
planted late if the regular pasture looks as if it would not last.
Canary grass is used in low places where clover or alfalfa would
drown out.

Among the newest of forage crops are Balboa rye, lespedeza,
Ladino clover, excellent for summer pasture where other crops burn
out, and kudzu, the prolific vine which is transforming much of the
eroded land of the South into green stretches again.

Science is even stepping in to determine the best time to cut hay.
In former years, farmers waited until blossoms were fully developed.
Experiments have shown, however, that much of the valuable protein of legumes like clover and alfalfa is lost by late cutting. Some
scientists place the loss of protein in late cutting as high as 60 per
cent, to say nothing of the greater palatability and tenderness of
early-cut hay.

Farm chores have yielded perhaps the least to the agricultural
revolution, but even here progress is being made. In the last few
years, researchers have discovered that "fast-milking" techniques,
both for machine and hand milking, produce more milk and cut
down on the likelihood of bovine diseases. One man with two machines can milk a cow every three minutes. Automatic silo unloaders
are on the market, ready to eliminate one of the greatest drudgeries
of farm work. Electric barn cleaners run a continuous chain of
paddles through the gutters, dumping the manure into the spreader
outside.

Electric motors grind and mix feed, operate hay and feed hoists,
power the big fans that cure hay in the barn out of the weather,
and run the power tools in the farm workshop.

Dairying has been made easier by the new milkhouse equipment.

Electric coolers for milk, and heaters for hot water to scald the milking utensils are now standard equipment on many farms.

Electricity has even stepped in to train cows, both in the field and barn. Electric fences, easy to string for temporary pasture, give the cows a gentle jolt and discourage "riding" fences to get at the grass on the other side. In the barn, the new "cow trainers" give Bossy another gentle nudge which soon teaches her to take one step backward and deliver the manure directly to the gutter, making the work of keeping her clean much easier.

Artificial insemination, scarcely heard of before 1935, now breeds literally hundreds of thousands of cows a year, and semen from purebred bulls is even delivered by specially designed parachute packages on regular airplane circuits. Besides reducing the spread of contagion within a herd, artificial insemination permits a farmer to build up a purebred herd from small beginnings, enables service from the best tested bulls, and permits breeding of heifers too light to support a mature herd sire. One more blessing, too little appreciated, is removal of a dangerous animal from the farm.

A chief reason why the agricultural revolution spread so fast is the fact that unlike competitive businesses, farmers delight in showing off their latest and best productive measures. A farmer who gets a big yield with a new strain of wheat or a special fertilizer combination brags about it to everyone who will listen, and practically begs the county agricultural agent to stage a field day at his farm, so that everybody can see the marvel.

One could wish that a similar revolution had been taking place in the thinking of farmers, but that would be overoptimism. It is true now, as it has always been true in the past, that a small segment of farmers keeps abreast of the latest developments, but far too many are still steeped in outmoded ideas, at least as regards soil conservation. The tenant farmer or even the owner who "mines" the land, extracting the richness in a few years of intensive cropping, and then moves on to greener fields, is still with us.

But such farmers are likely to diminish in numbers. A double incentive is working to save the land. One prong is made up of incentive payments from the government for wise soil conservation practices. In the last few years, farmers have been able to get a large proportion of their costs back, for outlays like terracing, liming, erosion and weed control. By and large the progressive farmers have jumped with both feet into adoption of these methods, and their example is having a powerful effect on their more backward neighbors.

The second stimulus is economic. With the homesteading areas vanished, there is no such thing as good cheap farmland anymore. At present prices for farmland and buildings (see Chap. 20, "Choosing the Land"), even the most grasping of farmers can't afford to despoil one farm and then move on to another. Perhaps for the first time in the history of this country, it is good business as well as good farming to conserve the soil. The farmer who conserves and enriches the topsoil of his land is the man who will reap the biggest and most profitable harvests.

For these two reasons, farmland in this country is growing more productive, instead of less so. The soil nutrients are being restored by barnyard manures and commercial fertilizers. Strip cropping, terracing, and contour plowing are saving the soil itself from being washed away. We are approaching—though in most places we have not yet reached—intensive farming practices.

-13-

WHERE THE GARDEN CAN
HELP MOST

Quality, quantity, and variety, on "the most productive acre" of the farm.

Try at least one new vegetable each year.

Start the garden in the house, while the ground is still covered with snow.

Electric hotbeds are easiest to make. Don't forget a chart, for transplanting later.

The "first-early" garden—a worth-while gamble with Old Man Weather.

Take extra time and care in planting seeds, or some of your other soil-fitting work will be wasted. Plant shallow in spring, deeper in midsummer.

Check maturity dates, to keep the crops from "bunching." Small, frequent plantings are best for everything but big canning crops.

Don't bother to cultivate more than enough to kill the weeds.

"Combination" plantings are only theoretically good.

Beginning the harvest. How to store vegetables.

Leave carrots and beets until after frost.

Sow winter rye in the bare spots, for green manure.

SINCE a garden is something which can be tailored so exactly to each family's tastes and wants, it would be foolish to set down a list of things which you must plant, with quantities and varieties specified. I have city friends who are as happy as larks, planting nothing but tomatoes and cucumbers—because they happen to have a passion for pickles and homemade tomato spaghetti sauce, and don't really care what else they eat from the garden. And I know other people who try to make their gardens what I myself think they should be—a year-round source of quality, quantity and variety for the home table.

I put quality first, because it is the one thing you can get nowhere else but in your own garden. You can plant the more delicate varieties of vegetables, since you don't need tough strains to stand the long hauls to market and on the grocer's shelves. You can make sure that the soil they grow in contains the necessary minerals and other constituents to produce the proper vitamins—for if these elements are not in the soil, you can bet they aren't going to be in the plant, as many persons fed on refrigerator-car vegetables are beginning to find out. And lastly, you can pick these vegetables when they are at perfection, and bring them really fresh to the table, rather than wilted and in a state of partial decomposition caused by several days of transit. Even the frozen food purveyors can only approximate this freshness and flavor, and not wholly achieve it. For a person accustomed to nothing but store-bought or canned vegetables, it may even be necessary to spend some time educating the palate to appreciate the fine nuances of truly fresh vegetables.

Quantity is something dependent only on the strength of your back and the length of your acres. I think it cannot be denied, however, that the garden is the most profitable acre on the farm— even when it is only half an acre. With good management, a garden will produce food for immediate use and for storage which would cost several hundred dollars if bought in the store. What other acre on the farm can make that boast? With a combination of quick freezer, canning and root cellar, a good-sized garden will keep you well supplied with vegetables throughout the entire year.

Variety, the third great argument for a garden, is not always so easy to achieve as quality and quantity, because the hazards of insects, weather and plant diseases may wipe out several special crops that you had set your heart on. Nevertheless, variety can be pretty well guaranteed by frequent plantings of sufficient varieties. And by variety I mean not only the different kinds of vegetables, but also the various strains among each kind. For instance, in one short row you can grow half a dozen kinds of leaf lettuce, such as black seeded Simpson, oak leaf, Paris Cos, Grand Rapids, and the like, which can be blended for a salad, or used individually, depending on what your whim of the moment dictates. Literally dozens of kinds of beans await your acquaintance, from the meaty concentration of French horticultural beans, the rare delicacy of small green limas, green string beans for canning, freezing, or boiling, or the slightly heartier flavor of pole beans, to the surpassing tenderness of really fresh wax beans, creamed and piping hot. Edible soy beans are good too.

One could range through the whole seed catalogue, outdoing the seed salesman himself in praise of this leek, that carrot, or this cantaloupe. The fun and profit of gardening, I think, is to try at least one new vegetable each year, or unusual strains of several old friends. Some will be suited to your soil, climate and palate, and thus will broaden the variety of your table. As for the others—you will perhaps say as I do, "Ah, next year I will have the perfect garden, for now I know at last just what to plant, and how to care for it." Possibly it is the feeling of having not only one, but several more chances, which makes gardening such a fascination, for those recurring chances do not always appear in other lines of endeavor.

The seed catalogues, arriving about the first of the year, with dismal winter still not half over, are the true harbingers of spring, full of enthusiastic promise. Now is the time to sit before the open fire, scanning the catalogues hastily, and marking all attractive possibilities, while your imagination creates the bumper crop of the ages. Later you can go back over the catalogues, weighing your choices, and comparing prices of one seed dealer with another, until it is time to put in the order.

Even before the snow is off the ground you can begin actual gardening. I like to start with a seeding of Pascal celery in a crock in the kitchen window. Celery is a slow germinator, and by the time the hotbed or cold frame is ready, the spindly celery plants will be ready for transplanting there, on their way to eventual setting out in the garden. Heat the dirt first in the oven, which will prevent "damping off" by mildew later. Scatter the seed fairly generously, and barely cover with soil, pressing down firmly. A glass on top will keep things moist for the first few days.

The hotbed or cold frame is next on the garden list, and it can be as simple or as elaborate as you like. Pick a place protected from the wind, with a south exposure, and out of the line of running children and dogs. The classical hotbed of our ancestors had concrete sides sunk four feet in the ground, and was filled with three feet of good horse manure, topped by half a foot of black earth. The decomposition of the manure furnished the heat for the hotbed, but this process was subject to wide variations. If the manure was soaked too heavily with water, it overheated, and burned up all the seedlings. And if it was old and dry, it sometimes did not furnish enough heat to cause the seeds to germinate.

You can still build the old-fashioned hotbed, using dirt sides and a wood frame at the top if you don't want to pour concrete, and spare storm sash for glass on top, instead of the regular hotbed sash, which is made of overlapping panes, to drain off the water.

Much simpler to make is an electric hotbed, with a wiring set which you can buy either from the seed dealer or in a hardware store. With an electric hotbed all you need is a box sitting on top the ground, banked from the outside. Thermostats control the temperature, and the cost of current consumed is not supposed to be excessive.

The hotbed, heated either by electricity or manure, gets the seeds started germinating, and carries them through the below-freezing nights of early spring. Later you can use a cold frame for additional planting, or, if you don't care too much about having the earliest things, you can skip the hotbed and just start with a cold frame, being content with vegetables that will be two or three weeks later than if raised in a hotbed.

Cold frames, and hotbeds of both the electric and manure types, are usually built with the front, or south, end about four to six inches lower than the back, and the soil, a foot below the glass, sloping in the same direction. Banking of dirt along the sides will save heat and prevent nipping by cold nights. For very cold nights, a covering of an old blanket or straw should be thrown over the glass. As spring advances, the frames are lifted a few inches each sunny day to prevent excessive heating.. About a week before the plants are to be set out, the frames are removed entirely, to give the plants a chance to harden.

Do not forget, in planting hotbed or cold frame, to draw a simple chart, showing the location of each variety. This is especially important, for instance, if you are trying out four or five different kinds of tomatoes, to find the ones best suited to your soil and taste. Make another chart when you transplant to the garden, so that you will actually know the varieties when it comes time for them to produce fruit. Annual or perennial flowers, early cabbage and cauliflower, tomatoes, celery, head lettuce, leeks and similar long-maturing vegetables are suitable for planting in the hotbed or cold frame.

By the time the first sprouts have pushed up in the hotbed, you will probably be able to stick in a "first-early" small garden in some corner where it will not interfere with plowing the regular garden. As soon as the frost is out of the ground and the land can be worked, I like to spade up a few square feet of soil and put in radishes, lettuce, a wide row of carrots, and a few beets. In one out of four years a late snow or cold snap may kill them all off, but it is a gamble worth taking for the sake of having fresh vegetables a whole month before the regular garden is producing. Even if the ground is really too sticky to work up well, such an early garden seems to me worth while.

After the frost is out of the ground, and the land has dried out somewhat, you can plow and fit the soil of the regular garden.

Plowing too long before you intend to plant merely produces an early crop of weeds before your seeds are even in the ground. And of course plowing too late means that all of your crops will be that much tardier. To tell whether the land is fit to work, make a small ball of earth in your hand and drop it. If it fails to break and scatter, the ground is still too wet. Ground worked when it is too wet is apt to "puddle," changing its composition so that it bakes instead of remaining porous. Land in such condition may take years to restore to normal soil, so those extra days and hours are not worth the chance you take.

After the land is plowed, it should be harrowed or disked almost immediately, to conserve the moisture. (See Chap. 15, "The Living Soil.") Then follows hand raking, to remove surface sticks and stones and leave a finely pulverized inch or two of topsoil as a seed-bed. Don't spare the rake at this point. A good seedbed promotes a uniform stand of crops, and can save you at least one cultivation. Remember, in marking the rows, that if you plan to use a power cultivator, such as horse, tractor or garden tractor, the rows should be straight, run the long way of the garden, and be of uniform width—at least three feet. If you are not using power equipment, make the rows straight anyway, to please the esthetic eyes of yourself and the neighbors, but the rows can be of varying widths, down to eighteen inches for small stuff.

If the garden has not been heavily fertilized before plowing (and a garden can profitably use from six to ten times as much fertilizer as is ordinarily applied to a field), you can use commercial fertilizer within the row, at the time of planting. I often make a small furrow with the hoe handle, scatter a dusting of commercial fertilizer along the furrow, and work it sketchily into the soil, to prevent burning the seeds. This extra "shot" gives the young plants a swift start, to get ahead of the weeds, and begins good root systems while the weather is still favorable for rapid growth.

Having spent several hours in picking out your seeds, and many more hours fitting the soil, don't speed up and get careless when it comes to the critical point of putting the seeds into the ground.

After all, that's what you've been doing all this work for, and if you don't plant your seeds right, your crop will be proportionately poor. The general rule is to cover the seeds with four times their own width of soil, but this should be modified by the time of year. In spring, when there is plenty of moisture in the soil and not too much heat from the sun, plant very shallow. Later on in the summer, when the ground is drier and the sun hot, put the seeds considerably deeper, so that the subsurface moisture can reach them, and the sun will not burn them out. After planting, firm the soil above the seeds, either by walking the length of the row or tamping with the hoe blade. This brings the seed firmly in contact with the soil and its moisture.

If you sow too thickly (as most people do), you will be spending a lot of time later thinning the plants. And if you sow too thinly, the stand will be ragged. A good plan is to sow a little more thickly than the stand of mature plants. This small extra percentage will take care of the seeds which do not germinate, plus the plants which fall prey later to the careless hoe and the eager insect.

Many seed catalogues have a paragraph or two on planting and cultivation hints for each vegetable, and you can be a pretty good gardener just by following these instructions, and those on the seed package. There are, in addition, a profusion of books on every aspect of gardening, and stacks of pamphlets in the county agent's office.

What can you plant first, and how much of each kind? In general, everything but beans and corn, which must wait until the ground is really warm, can go in right away. This includes peas, beets, carrots, radishes, lettuce, chard, spinach, parsnips, onions, potatoes and the like. Some of them, like lettuce and parsnips, need cool weather to germinate, and will not sprout when the temperature averages above 60. You can fool the lettuce later in the season by "planting it in the icebox" first. Moisten a couple of blotters, sprinkle the seeds on them, put the blotters together in an envelope, and leave them in the icebox for three days. Then plant in the garden, water heavily for the first few days, and you should have nearly 100 per cent germination. The three days of icebox treatment are enough to start the lettuce germinating, and sun and moisture will do the rest.

Some seeds, like those of carrots and lettuce, are so small that it is difficult to sow them thinly. If you are sowing by hand, mix the seeds with finely pulverized dirt, and scatter in the row, to get an even stand. There are also hand-powered garden cultivators with seeding attachments, which distribute the seeds evenly. For large gardens they are worth the investment.

Don't forget the maturity dates when you are planting. The seed catalogue will give a maturing date for each variety, from planting or transplant time to harvest. Thus you can plant "succession peas," usually three varieties which mature at succeeding intervals, all at one time, but they will ripen at different times. If you are meditating a heavy canning schedule, pick varieties of beets and other vegetables which will be ready for canning at some other time than when you are sweating over the preservation of peaches and pears, for instance.

Make large plantings at one time of the vegetables you expect to can or freeze, but keep the other plantings small and frequent, to insure constant variety throughout the summer. Instead of one long row of radishes, for instance, which would be at their best for just about one week and then go to waste the rest of the summer, make a small planting every week. I usually pick a spot at the end of a row of some other vegetable, or in the middle, where the plants failed to germinate. Each Sunday I scratch an area about a foot square with the rake, scatter the seed broadcast, throw on a couple of handfuls of dirt, step on it, and pay no more attention to them until they are ready to eat in about three weeks. Beets, beans and lettuce are other vegetables suitable for frequent planting.

Small, frequent plantings also enable you to get two crops off the same land. Thus early peas or beets can be followed in the same spot by a late planting of beans, or vice versa. A sprinkling of commercial fertilizer, to replace the nutrients taken from the soil by the first crop, will make the second crop do better. Some vegetables, of course, such as broccoli and chard, keep on producing all summer, and hence do not need to be replaced. Others have such a long growing season that there is time for only one crop. Parsnips, potatoes, seed onions, corn and tomatoes fall into this class.

At about the time you make your first big planting in the regular garden you should also be preparing the seedlings which will be transplanted with warmer weather. The small celery seedlings in the house, and the tomatoes in the hotbed can be transplanted to a cold frame, to help them develop a better root structure and make them less crowded. If you want early melons or squash, dig up chunks of sod about four inches square, turn them upside down in the cold frame, and plant your melon or squash seeds in them. These plants are ordinarily too delicate to stand transplanting, but the entire sod can be picked up and moved to the field as soon as the plants have their second leaves, without giving them too much of a setback. This method will give you melons a month earlier than your neighbor. In northern climates it is the only way of being certain of melons before frost.

While you are waiting around for the first plantings to show above ground, weed the strawberry bed after a good rain, and cultivate around the raspberries and other fruit bushes. You can also experiment in a corner of your garden with some of the new chemical weed killers, applied to the soil before vegetables appear.

When the first vegetables are up, it is usually warm enough to put in beans and corn. You can plant successions of sweet corn by noting the maturity dates, and allowing about ten days between each, so that they will not cross-pollinate. Even though they do not taste quite so good as the longer-maturing varieties, I like a few rows of quick-maturing corn, so as to be able to get at eating it as soon as possible. Keep on planting corn at ten-day intervals, so that it will last the summer. As the season advances, go back to the quick-maturing varieties for your last planting, which should be put into the ground so that it will mature just about the time of the average date of first frost for your area. Remember that corn pollen drops from the tassel to the silk, and the pollination is affected by the wind. For this reason several short rows in a bunch will get good pollination, while one long row will not. Keep your sweet corn at least forty feet away from any plantings of field corn, or from the popcorn you put in for the children, or you will be chomping on many a tough ear.

You can set out the delicate items from the hotbed, like tomatoes and melons, under hotcaps if there is still danger of frost, or you can wait until all danger of frost is over, and set them out in the open field. On transplants do not put commercial fertilizer in the hole or it will burn the roots and kill the plant. You can rake some in later, near the surface.

As soon as the plants in your garden are up sufficiently to mark the rows, get busy with the hoe or the power cultivator. Frequent shallow cultivations—just scratching the surface of the ground—are better than deep digging. The shallow cultivation kills any weeds near the surface. Deep cultivation, on the other hand, is likely to dry out the soil, and also brings weed seeds close enough to the surface to germinate, instead of leaving them dormant. Don't try to get too close to the roots. You will do more harm than good.

If you tire of wielding a hoe, you can try the mulch method of gardening. Lay down strips of building paper between the rows, or cover the space with a six-inch layer of straw or hay, and you won't need to hoe. The mulch will keep down the weeds and keep the soil moisture up close to the surface.

Fortunately, a garden which is planted early gets a head start on weeds and insects, but as the summer advances you will have to be ever vigilant with the spray gun and duster. You get better results if you spray and dust at the first signs of infestation, rather than waiting until the insects have made lacework of the leaves and laid a bunch of eggs to cause you trouble next year. Because of the rapid development of new spray equipment and sprays, no attempt is made here to prescribe sprays for the various insect pests. Your seed store or seed catalogue will be equipped with the latest information.

Try to work with the hot weather of full summer, rather than against it. Cauliflower, for instance, simply will not head out in hot weather. Get your plants in as soon as possible, for a crop before the hot weather, and start seedlings the middle of summer for a fall crop when the weather turns cool again. Leaf lettuce, if you can get it to germinate, does fairly well in hot weather, but many kinds of head lettuce will bolt to seed unless you make a lattice rack above them to cut out part of the sun. There are some varieties, such as Bibb lettuce, which do well in hot weather.

In the midst of summer, corn, beans, beets, carrots and onions thrive, so plan to have your big crops of those vegetables mature at that time. As soon as you have three squashes set on each runner, pinch off the new blossoms and pinch the ends of the runners, and you will get bigger squashes. If root borers are getting at the vines, throw a couple of shovelfuls of dirt on top the hill, and put a heap of dirt on the center of each runner, which will then take root in the middle. Because of the unfortunate propensity of some truck gardeners in picking squash while still green and foisting them on an unsuspecting public, many persons have never tasted a truly ripe squash. For the green varieties, such as Hubbard, buttercup and table queen or acorn, the bottom, where the squash rests on the soil, should show a definite orange. If the color is merely yellow, the squash is not yet ripe. You will get about the same weight of squash per acre, no matter which variety you plant. The acorn or table queen has a very delicate flavor (which is too flat for me to enjoy) but because of its thin skin will not keep far into the winter. It has the advantage of being a handy serving size. For a richer flavor, I much prefer the buttercup squash, which grows between eight inches and a foot in diameter, and one of which will make a meal for a family of five. Buttercup squash will also keep well past the first of the year. Few families are big enough to tackle a whole Hubbard squash, but it can be cut in half, and the unused portion covered with wax paper, for eating a few days later. Because of their thick hide, Hubbards will keep well into March and are worth planting for that reason. I confess I also relish the sense of opulence a big Hubbard gives on the dinner table. Buttercup and Hubbard have a similar flavor.

If you plant pumpkins, either for Halloween merriment for the children or for eventual pies, keep them twenty to thirty feet away from the squash, or the bees may produce some crosses.

Some gardeners point out that squashes or pumpkins can occupy the same land as corn, but if you have enough space for separated plantings, you are likely to get a better crop. It is true that the vines will run among the corn, but if there is just enough moisture and nourishment for one crop, it means that neither will attain full size if they are planted together. An alternative is to skip two or three hills of corn each time you put in a hill of squash or pumpkin. Or plant where the corn failed to come up. Of course, if you have a very early variety of sweet corn, where the stalks hardly grow to five feet in height, you can plant corn or squash with them, and both will do well, because the corn will not take much from the soil, and will be withered and gone before the vines have really begun to develop much fruit. A disadvantage of mixing corn and vines is that one is likely to tramp on or tear the vines while harvesting the corn.

Another combination often suggested is pole beans and corn. Here again the corn, with its early start, is likely to steal all the moisture and nutriment, leaving very sickly beans to be harvested. One of the easiest methods for pole beans, either green or lima, is to plant them sparsely in a row, and then stretch an old section of woven wire fencing, about five feet high, vertically above them. When frost comes and the vines wither, the wire can be rolled up, doused with a pint of kerosene, and the vines burned off in a jiffy. If you don't want to burn the vines off, the wire can be stretched near the driveway, and will make a fairly passable snow fence if the vines are thick.

As the summer wears on, the harvest for winter use should begin. Extra strawberries, raspberries, and asparagus can be frozen or canned. (See Chap. 14, "Cash Crops and Truck Farming," for directions for raising.) Depending on freezer space and the ability of the housewife to can, you should be able to get virtually all your vegetables for the winter, plus the root crops which can be stored in the cellar. For a family of five, a total of 400 to 450 quarts, in freezer and on the shelf, would not be too much. Unless the freezer has more than twenty-five cubic feet of capacity, most of the vegetables will have to be canned, if this total is to be reached. One easy way is to freeze small quantities of the choicest vegetables, two or three quarts at a time when they are in excess supply, and can the large crops when they ripen in mass quantities.

The earliest root crop to demand storage will probably be onions. Grown from seed planted early, and thinned to four inches apart in the row, they will just about mature by frost, but if you plant onion sets or plants, they will mature by the middle of summer. The time to harvest is when the tops wither and die down, but it is not necessary, as our ancestors thought, to "barrel" the tops down to hasten maturity. In fact, rolling a barrel over the tops will not speed up maturity. After the tops are down the onions should be dug out of the ground, or a good rain may start them growing again, producing new flesh which will cause them to rot when stored.

After they have been turned out on the ground for a couple of days to dry out, the onions should be collected in well-ventilated shallow crates or small net bags, and stored in a cool, dry place. Usually the garage will be most desirable, until freezing weather, because of the tendency of onions to embalm the air of any place they are stored.

Except for those needed for immediate use, potatoes are best left in the ground until the end of summer. Digging them earlier causes them to dry out, and thickens the skin. You can retard this process by leaving them in the ground as long as possible. If the vines fail to wither, tear them up or kill them chemically some weeks before digging, and the potatoes will store better.

Beets can be harvested at frost time, but like carrots are the better for a nip of cold, which "sets" the cells of the plant. Because beets "bleed" and dry out if the leaf stalks are cut close to the bulb, leave about three inches of stalk when trimming them, and store them in the root cellar either covered with sand or in a crock which has a moistened cloth cover.

Carrots should be left until after frost, so that the chill will set the cells. Dig them before the ground freezes, cut off the top inch of the root to prevent the green of the leaf stalk from rotting, and put them either in sand or in a crock like the beets. Both carrots and beets should be stored in the coolest part of the root cellar, preferably at about 38 degrees, and should be kept moist to prevent their drying out.

Squashes, on the other hand, need a high temperature, about 60 degrees without moisture, and should be stored so they do not touch each other. They should be harvested just before frost. Late cabbages can be hung from the ceiling by their roots, and will keep a little better if wrapped in paper. Potatoes keep best at about 43 degrees. If kept in a warmer place, and with too much moisture, they are likely to sprout. (A thermometer moved to various locations in the root cellar will give you a pretty good idea of what the temperatures are in each section. Remember, however, that stored vegetables throw off a little heat.)

There is some satisfaction, also, in checkmating the frost for a few vegetables. If you have a few hills of melons that look promising, bed them down with straw when the weatherman gives a frost warning. In the same way, cornstalks can be laid alongside celery plants to cover and incidentally blanch them. Tomatoes must also be covered if there is frost danger, or the vines will die. Old bags, blankets or rugs, as well as straw, can be used for the purpose.

Another alternative, if frost and really cold weather look likely, is to move the tender plants to the root cellar. Tomato vines, with the green fruit still on them, can be strung in the garage or root cellar, and you will have gradually ripening tomatoes for about three weeks. Celery can be transplanted to sand in the root cellar, and will keep fresh for several weeks, as it blanches.

When all the edibles are safely removed, you can collect and burn the remaining old vines, to kill any insect or bacterial pests they may harbor, and the garden is ready for winter, and heavy applications of organic fertilizer. If you have planned your plantings so that there are large bare spaces, it is possible in early September to sow winter rye, to be plowed under next spring as green manure, giving added fertilizer and humus to the soil. In small gardens you can sow the rye in bare spots after the middle of summer, either spading it under late in the fall, or waiting until spring.

ADDITIONAL READING MATERIAL: (free from U.S. Dept. of Agriculture) *Mint Farming*, F 1988; *Hotbeds and Coldframes*, F 1743; pamphlets on culture of each vegetable.

Asparagus Production, by W. B. Ward and N. K. Ellis, Purdue University Agricultural Extension Service.

From Supt. of Documents: *Making Fermented Pickles*, 5¢ FB 1438; *Asparagus Culture*, 10¢, Cat. No. A 1.9:1646; *Snap Beans for Marketing, Canning and Freezing*, 5¢, Cat. No. A 1.9:1915; *Blueberry Growing*, 10¢, Cat. No. A 1.9:1951; *Corn Culture*, 10¢ Cat. No. A 1.9:1714; *Cucumber Growing*, 10¢, Cat. No. A 1.9:1563; *Home Storage of Vegetables and Fruits*, 10¢, Cat. No. A 1.9:1939; *Insecticides and Equipment for Controlling Insects on Fruits and Vegetables*, 15¢, Cat. No. A 1.38:526; *Vegetable Gardener's Handbook on Insects and Diseases*, 15¢, Cat. No. A 1.38:-605; *The Farm Garden*, 15¢, Cat. No. A 1.9:1673; *Nutriculture* (hydroponics, or growing plants in liquid, without soil), 5¢, Cat. No. A 1.35:219; *Grapes for Different Regions*, 10¢ Cat. No. A 1.9:1936; *Savory Herbs, Culture and Use*, 10¢, Cat. No. A 1.9: 1977; *Growing Peas for Canning and Freezing*, 5¢, Cat. No. A 1.9:1920, *Raspberry Culture*, 10¢, Cat. No. A 1.9:887; *Strawberry Culture—Eastern U.S.*, 15¢, Cat. No. A 1.9:1028; *South Atlantic and Gulf Coast*, 15¢, Cat. No. A 1.9:1026; *Western U.S.*, 10¢, Cat. No. A 1.9:1027.

-14-

CASH CROPS AND TRUCK
FARMING

The extra income in truck farming comes out of your back and
fingers.
How to raise cash crops without touching the land.
Peas and corn for the cannery.
Don't try truck farming unless you are certain of a market.
Produce small crops early for high prices, or bumper crops later for
quantity prices.
Melons and squash take little work.
Asparagus, strawberries and raspberries mean a heavy harvesting job.
As long as women love to put up jelly, you can sell gooseberries and
currants.
Other cash crops.

THE cash crops you can raise in the country offer entrancing vistas
of added income in theory, and really work out that way in ac-
tuality—if you can stand the pace. Remember that intensive production, such as in raising strawberries, melons, tomatoes, or similar
truck farming crops, means many long hours of picking and preparing for market, to say nothing of the time spent in raising the fruits
and vegetables to the picking stage. The extra money per acre is
there—but it is in your fingers, your time, and your back, much
more than it is in the land.

However, if you shudder at the prospect of long hours filling strawberry crates in the hot sun, there is an easier way. You won't make as much money as with the intensive crops, but for the time and investment you'll do pretty well. The easiest of these lazy man's cash crops is hay. Once your land is seeded, you can sell the standing crop for someone else to harvest, at so much per acre, load or bale. And you can keep right on doing it, year after year, so long as the stand is productive and you keep up the fertilizer. If you have planted a legume for hay, such as alfalfa, you will actually be enriching the soil at the same time, by building up a deposit of nitrogen. A variation would be to sell the first cutting for hay, and let the second mature for seed, to be custom-combined. Both these possibilities, of course, presuppose that you are farming in a neighborhood where extra hay is wanted.

Other crops which can be turned into cash with no expenditure of effort on your part are raising corn or small grains on shares. Under this plan the sharecropper fits the land, puts in and harvests the crop, and delivers your share wherever you specify. If you don't want to be bothered with storage to await higher prices, he can deliver your share direct to the mill, and all you have to do is collect the check. Such sharecropping can be done in virtually every part of the country where general farming is carried on.

You can get a greater return per acre, short of intensive truck farming, if you try certain other crops, which call for a little work on your part. They also differ from hay and sharecropping by requiring a near-by available market. Unless that market is available, and you have a contract to sell, don't plant. One such crop, for instance, is peas or corn for canning. If you plant two or three acres of peas, the cannery will tell you what day to put them in, and what day, and even hour, to harvest them, and you must be prepared with machinery and time for that harvest. Raising sweet corn for the cannery requires even more work, for you must not only plant, but also cultivate, and the drain on the land is more severe. It is also possible to rent several acres to the cannery, letting them put in and harvest the crop.

But suppose you want to make a killing—without completely killing yourself. The methods previously outlined will bring in a fair income, but nothing spectacular. If you want to get into the production of crops that will produce in the hundreds instead of scores of dollars per acre, the only method is truck farming. An absolute essential, of course, is a near-by market, preferably without too many competitors. It can be a large or medium-sized city, a canning or freezing plant, or even a railroad station heading toward a big town some fifty miles away. You need to inquire about kinds of vegetables, and particular strains of those vegetables, which will find favor with the buyer, or your prospective profits will just be whopping losses.

The truck farmer has two choices of how he can conduct his business. He can be first in the market with quality produce, or he can specialize in quantity, making smaller margins but turning out a lot more stuff. Protected southern slopes, for instance, coupled with extra care with the hotbeds and in transplanting, might bring him into the market with tomatoes two weeks ahead of his neighbors, when the price is three times as high. This premium on being the early bird does not apply to canneries and quick-freeze plants, but it does in the vegetable and fruit markets. Marketing through a growers' co-operative, with the advice and help that such an organization can give, is another possibility. (See Chap. 25, "Marketing, Co-operatives, etc.")

The man who plans to be first in the market needs good, productive land, sheltered from adverse winds, and he will be spreading his work over a longer period of time, than will the man who merely produces for quantity. With luck and skill the first-early producer may make considerably more.

Whether you plant early or plant for quantity, full-time truck farming is a highly specialized occupation, which requires a strong back and very long hours of manual labor during spring, summer and fall. Because it is so highly specialized, there is no room here to set forth its complexities, but there are pamphlets available on each of the specialized crops, to which the reader is referred.

Now, how about an acre or two of cash crops (other than staples like corn or soybeans) for the part-time farmer, the subsistence man, the week-end farmer, or even the full-time farmer who could use a couple of hundred dollars of extra cash in the middle of the summer? Here again, near-by markets are the deciding factor. If you are many a mile away from anybody who will buy your produce, there's no use in planting it. Suppose, however, that you are within a few miles of a good-sized village or small town, and a couple of storekeepers tell you, in the indifferent way storekeepers have with farmers, that they might buy your stuff.

Your choice of a cash crop will depend somewhat on your available manpower, and how quickly you want the cash. Sweet corn, melons and squash are three crops that can be raised and sold the same year. If you have a job in town, it is easy to pick a few bags full of corn and deliver them on your way to work. And you may even be able to dispose of some surplus from your garden, such as beets, carrots and radishes. For years, by planting a few extra rows of corn, I paid for all my garden seeds and house shrubs with corn from a few extra rows. In addition to several smaller plantings to come on during the entire summer, I made two big plantings, one to come on very early, and the other later. Aside from what we ate ourselves, which usually turned out to be an enormous quantity, I sold all the early corn to stores when the price was still high. Later, when the price dropped, we canned the surplus from the second big planting. This corn business was a small matter, to be sure, involving a few hundred dozen ears, but it made our garden produce absolutely free, except for the work involved.

Melons and cantaloupes are another crop where one man can do all the planting, care, and harvesting. You can get a head start by planting the seed in inverted pieces of sod in a cold frame, transplanting later when the weather is warm. In northern areas, this is the only way you can get melons. (See Chap. 13, "Where the Garden Can Help Most.") A shovelful of well-rotted manure in each hill, dusting with powder to kill the borers, a little early attention to weeds, and you can sit back and wait for the harvest. Melons do best on a light sandy loam.

Like melons, squash require very little work. Since they are sold for fall and winter use, being first in the market is no particular advantage, as it is with corn and melons. If you have a conscience, you will plant early enough so that the squash actually mature (the spot in contact with the ground shows orange when they are fully ripe, rather than merely yellow), even if the storekeeper and the customer don't know enough to buy ripe squashes. You can easily raise five or six tons of squash per acre, but be sure to find out first where you can sell them, and what varieties are desired. Some districts favor the small table queen or acorn squash, while others like the big Hubbards and the somewhat smaller buttercup squash.

Some other cash crops require two or three years of work before you collect any money, and a few need plenty of hands for a short, heavy harvest season. One of the easiest is asparagus, but it takes three years for a bed to begin to produce, after which it will continue for as long as fifty years. Our forefathers laboriously dug trenches nearly two feet deep, filled them with manure, then soil, and then set out the asparagus plants. Recently someone took a good look at the roots, noticed that they always spread out horizontally a few inches below the surface, and concluded that a subsoil of fertilizer was largely wasted. The present method is to set one- or two-year-old plants half a foot below the ground, a foot apart in rows four feet apart, and to pile the fertilizer on top, instead of beneath the plants. Asparagus is a heavy eater of fertilizer, and each year's crop is directly dependent on the quantity of fertilizer you spade into the soil in early spring. Because the maturing season lasts nearly two months, it is a crop that one man can handle.

The berry crops are in a class by themselves when it comes to harvesting. A strawberry bed set out early one spring will put forth runners for new plants during the summer, but all blossoms should be pinched off, unless a light late crop is desired. The next year the bed will produce heavily, with the picking concentrated in a two-week period. This is when you need extra hands to fill the crates, and unless you have enough members in your own family willing and able to do this huge job of picking, your profits will be materially cut down by hiring help. The third year the strawberry bed produces slightly less heavily than the second year, and the fourth year, due to infestation with weeds, crowding of the plants, etc., it is no good at all. Hence a strawberry producer must set out a new bed every third year, and may be wise to set one out every other year. Because of the wide variety in size and flavor of berries, and the suitability of plants to various localities, consult your local agricultural agent for pamphlets on recommended varieties for your region, as well as for directions on culture.

Like strawberries, raspberries take a lot of hands for the picking, but since they mature later than strawberries, the same crew can be used. Raspberries are planted by sticking canes (two-foot sections of last year's growth) into the ground a few inches, a cane every two feet, in rows about six feet apart. These canes, planted in the spring, take root, and the roots will send up new shoots, filling the row.

The blossoms and berries come on last year's growth, so the second year, with the roots just getting established, the crop will be rather light. By the third year, however, the bed should be producing well. Each spring the gardener cuts or pulls out the deadwood which produced a crop the year before, cuts down the new growth of last year to about four feet in height to promote heavier setting and larger berries, and cultivates along the sides of the row, to keep the bushes from spreading into a jungle. With fertilizer, cultivation, and removal of deadwood, the bushes will keep renewing themselves for several years. Because raspberries are subject to several plant diseases, consult your county agent or agricultural college for varieties resistant to the diseases prevalent in your locality.

For all these cash crops which require much hand work and an acquired knack in raising them, the beginner is advised to go slow when starting out. Commence with a small planting, to learn how to handle the crop and how much you are able to do, rather than plunging into two or three acres of headache all at once. I have heard of families that were almost broken up because father calculated the price of a bushel of cucumbers, figured that an acre would produce a great many bushels, and then kept the children picking all summer until they were about ready to kill him.

Besides the big cash crops, consider also some of the supplementary fruits which can be a continuing source of small but steady income year after year. A row of gooseberry bushes along a fence, for instance, will fill space which ordinarily goes to waste, and the bushes are disease-resistant and practically indestructible. Currants need more space, and are afflicted with more pests and diseases. However, as long as women like to can jelly and make pies, these two fruits will always be in demand.

Another perennial which requires little care is rhubarb. A few roots, with a couple of forkfuls of manure spaded in around them, will produce heavily, and the roots can be sliced in two every third year to set out more plants.

Although the previous discussion covers the major basic cash crops, there is an endless variety of others with which you can experiment. Cucumbers, eggplant, peppers, cabbage, cauliflower, celery and head lettuce are just a few of the things you can raise which require some degree of special care, in addition to the staple root, leaf and pod crops. Since soils are so varied, and individual skills not uniform, here again the gardener should feel his way slowly, to learn the kind of crops that do best on his ground, and that he has the skill to produce in quantity.

And again it should be stressed that markets are all-important. Make reasonably sure that you can sell your crop before you plant it, or seed, labor and land investment will chalk up a loss instead of a profit. A little experimenting will soon show you whether you have the patience and strength for this most exacting of all tillage of the soil.

ADDITIONAL READING MATERIAL: See pamphlets listed at end of Chap. 13.

-15-

THE LIVING SOIL

Good, living soil is a balance of bacteria, chemicals and texture.

Billions of organisms live on each grain of soil, manufacturing and storing plant foods. But they need a proper balance of chemicals to do an effective job.

A simple, easy soil test tells you what you have in the soil, and what you need to add.

Plenty of barnyard manure is the best and cheapest fertilizer. How to preserve its soil-building values.

"Trace" minerals—small but very important.

Plowing under green manures adds both fertilizer and humus.

Plowing when the soil is too wet may "puddle" it for years.

A "perfect seedbed" is good for gardens, poor for field crops.

Rotating crops gives the soil a chance to recover. Two sample rotations.

Cultivation is solely to kill weeds, not to "preserve moisture."

Three types of soil erosion, and control methods for them.

"EVERYBODY knows what soil is—all you have to do is look at the land and you can tell whether it is rich or poor."

True enough, if you can see crops growing, preferably for two or three years in succession, and note whether the stands are heavy and well distributed throughout the field. Nevertheless, that eight to twelve inches of topsoil which has enabled Americans to live like kings is a complex of teeming organisms and mixture of chemicals. When in proper balance and tillage, it is a superb production factory. If one or more of the vital ingredients is lacking, or the land is mismanaged, the output will be meager.

Since every state has literally hundreds of different types of soil, and even the same farm may contain several varieties, there is no point here in sketching anything but the broadest outlines. Your county agent has an excellent soil map for your area, with a description of the characteristics of each type of soil to be found on your farm, plus its exact location in the field. The soils range from impoverished quick-drying sands and sandy loams through the rich silt and clay loams to the heavy clays. But these classifications merely tell you of their water-retention ability, how to handle them in plowing and tilling, and give little indication of the minerals they contain.

Much more important for your purpose are the microscopic organisms living on the grains of soil, breaking down the minerals and organic compounds into plant foods that the roots can use for nourishment. And equally important are the reserves of available minerals and fertilizers in the soil. The chief requirements for plant growth are nitrogen, phosphorus, potassium and calcium, all of which must be present in sufficient quantities, or you will have sickly, stunted plants. Each mineral is necessary, in order to get full benefit from the other minerals.

Soil experts sometimes use the illustration of a barrel to point out the importance of having each mineral in sufficient quantity. If each mineral is likened to a stave in the barrel, and the staves are of different lengths, the barrel can be filled only to the level of the shortest stave. An automobile assembly line could serve as another illustration. If you have plenty of bodies and motors, but only a few sets of wheels, you won't turn out many cars. In the same way, if just one mineral is deficient, a superabundance of the others won't do you any good.

Fortunately, all of the mineral deficiencies can be corrected by applications of the proper kinds of commercial fertilizer, in conjunction with barnyard and green manures. There is just one sure way to find out what the soil needs to get into perfect balance, and that is by a soil test, simple and easy to perform.

A spade and a few clean quart jars or paper bags are all you need for the soil test. Make a vertical cut about six inches deep with the spade, and then shave off a thin peeling of soil from the vertical face of the cut. If the soil is of the same general character, you may want to make three or four such cuts in different parts of the same field, and mix all the samples thoroughly in one jar. Use a separate jar for each field, number them in whatever way will enable you to identify the field, and take the jars to your county agent. He will either test them or tell you where the tests can be made. It is a good idea to note with each jar the kind of crop you intend to plant in that field next year. This will enable the county agent to advise you on the quantity and kind of fertilizer best suited for that field and crop.

While commercial fertilizers will enable you to restore mineral deficiencies which may have existed for years, your cheapest and most effective fertilizer will be a barnyard manure. For crops like corn, which is a voracious consumer of soil fertility, you may need to supplement with small amounts of commercial fertilizer, but plenty of manure will do the land the most good. Not only does it provide vast quantities of nitrogen, which is the biggest plant food requirement, but it is rich in amino acids and various complex chemical compounds not yet fully understood by scientists, but which for centuries have demonstrated their wonder-working properties in the soil. The true soil enthusiast, in fact, becomes virtually lyrical when he starts to sing the merits of barnyard manure as a soil builder. Along with its variety of chemicals, the manure also furnishes organic material in the form of decaying vegetable matter, immensely multiplying the activity of the soil bacteria which, by the billions per square inch, change the minerals and organic compounds in the soil into usable plant nourishment. The addition of good manure acts like throwing the power switch in a factory, starting things humming where before all was quiet.

Unfortunately, barnyard manure can lose half its value if it is not handled properly. If it is piled outside, exposed to the weather, the rains are likely to leach out many of its valuable constituents, particularly the nitrogen-bearing ammonias in the soaked bedding. Great losses in fertility also occur when the manure is spread immediately during the winter on sloping fields, subject to heavy runoff of water.

One simple chemical step will do much to preserve the fertilizer value of the manure as soon as it is produced. This is to sprinkle superphosphate on the manure and moist bedding. The superphosphate combines chemically to "fix" the nitrogen in the ammonia, so that it will not later evaporate when exposed to air. Use about one pound per cow per day.

The next important step is either to get the manure spread at once on the land, or to store it in a concrete-lined pit, sheltered from the rain, until it is ready to be spread. This storage, if the manure is kept moist but not soaked, will preserve nearly all of the fertilizer value.

Soldiers who were in France during the First World War never tired of telling how the French peasants stored their barnyard manure in pits near the house, and pumped the liquid (usually just about lunch time) into pails to carry to the fields. Esthetically the practice may have been bad, but agriculturally it was perfect, since most of the fertilizer value is in the liquids.

In addition to barnyard manure and commercial fertilizers, the soil may also be deficient in the so-called "trace" minerals, the importance of which has only come to be recognized in recent years. The most important of these are boron and magnesium. Their quantity, compared to the total amount of soil nutrients, is ridiculously small, but like the yeast which starts bread, if they are missing the product falls flat. As little as forty pounds of borax per acre, for instance, is enough to correct a boron deficiency.

Our ancestors, with the relatively low-producing seeds available to them, got by fairly well with barnyard manure and an occasional application of marl or lime for their fields, but that sort of practice won't maintain fertility today. The new hybrid corns, and the greatly increased productivity of new strains of plants in the small-grains family, take more out of the soil than the farmer can put back with manure from the stock the farm is able to carry.

Some of this gap can be taken up with commercial fertilizer, but another cheap and rich possibility is green manure. Crops like winter rye or sweetclover can be plowed under in the spring, adding humus and nitrogen respectively to the soil. (Because of its extreme hardiness, new sweetclover seedings should not be plowed under in the fall, because the root crowns will remain alive, and become a serious weed menace for the next year's crop. To avoid this danger, the clover should be plowed under in the spring, when the stalks are three or four inches high.)

Like every other living thing, the soil responds to good care and suffers under bad care. Thus plowing at the wrong time, or cultivation when too wet, can so change the texture of the soil that it may take several years to bring it back into good shape. This is especially true of the heavy clays and clay silt loams. Unlike the sandy type of soils, which can be worked at almost any stage of moisture, soils with a large clay base will pack into big, hard lumps if plowed or cultivated when they are too wet. The lumps then cake and dry out, and are almost impossible to break up. Working clayey soils when they are too wet also is likely to "puddle" them so that the soil is a solid mass instead of separate grains.

An easy way to tell whether the land is too wet to plow or cultivate is to make a loose ball of dirt and drop it. If it fails to break into several small pieces, the land is too wet to work. The "glisten" of the furrows as they are turned over by the moldboard of the plow is another indicator. If the plow sole leaves a definite track of moisture in the furrow, stop plowing. (Incidentally, don't plow or cultivate with valuables or important papers in your back pockets. Many a farmer, hopping on and off his tractor seat to adjust the plow after stones, has plowed under a billfold of money and never realized it until the field was done.)

While the light, sandy soils can usually be worked almost any time in the spring, the heavy clay soils are usually plowed in the fall, when weather conditions are better. This also exposes the clods to breaking-up action during the winter, and helps to accumulate and conserve soil moisture. Sod, such as in a pasture or hay field, is also frequently turned in the fall, both to rot the sod, and to expose grubs and worms. Huge flocks of blackbirds are especially valuable at this time in cleaning up the white grubs which later will become June bugs.

Some beginning farmers make the mistake of trying to obtain a "perfect seedbed," which is good in a garden, but does not produce the best crops in the field. A finely pulverized topsoil, almost like dust, looks wonderful, but it may pack at the first rain, and make a crust that the seedlings cannot easily pierce. A loose, crumbly surface is best. So much depends on the type of the soil and the degree of moisture that no general rule can be laid down for "fitting" the soil after plowing. If you are in doubt, ask your neighbor. Often sods which have been turned in the fall or spring are disked to slice them up, and the land is then dragged with a tooth harrow, to firm and level it. If a field is badly infested with quack grass, a spring-tooth harrow can be used each direction of the field, to tear it up and get the roots on top, before harrowing.

The matter of seedbed preparation is bound up with crop rotations. Tests at several U.S. agricultural experiment stations have confirmed the long-time wisdom of farmers who held that land on which corn is to be planted must be plowed in order to obtain the highest yield. On the other hand, where soil and other conditions are right, oats and other small grains do better if the land is merely disked and harrowed instead of plowed. The small grains, also, do better on land which raised a cultivated crop the previous year, rather than on freshly turned sod.

That is one good reason for the three-, four- or five-year rotation of crops which most farmers follow. Other reasons are that rotation, with crops that take different quantities of nutrients from the soil, gives the soil bacteria a longer chance to break down the necessary minerals and plant foods into available nourishment, and to acquire other nutrients. Thus corn, planted in the same field year after year, soon exhausts the nitrogen supply, but if it follows a nitrogen-storing crop, like clover or alfalfa, it does not exhaust the soil, but merely draws on the nitrogen "banked" by the legume.

While a soil may be rich in chemicals and minerals, only a small proportion of this total is usually available in any one growing season in the form that plants can use. The tiny, ultramicroscopic colloids present on each particle of ground are the workers which break down the chemicals into forms soluble in water, which is the only way they can be absorbed by the plant roots. Some of the nutrients are converted into compounds insoluble in water, thus preserving them from being leached away, and saving them for later use when the chemical conditions are right.

There is no particular mystery about crop rotation, nor is it an inflexible procedure which must be followed year after year. In fact, it has to be flexible, to allow for things like drought, which could kill out new seedings and force the farmer to substitute an emergency forage crop like soybeans or Sudan grass.

The number of years in the rotation depends somewhat on the feed requirements, and on the nature of the land. On level lands, for instance, a three-year rotation is practical, while on rolling land or steep slopes a five- or even six-year rotation may be desirable, to keep the land in sod most of the time, and thus prevent erosion. The rotation is like the childhood rounds which you used to sing. One crop will start in the first field one year, and so on ad infinitum.

Suppose your fields are level and fertile enough for a three-year rotation, and you have decided your major crops will be corn, red clover and oats. You can have three large fields of equal size, or of course three groups of smaller fields. On paper, the rotation would look this:

	1st Year	2nd Year	3rd Year
Field 1	corn	oats	red clover
Field 2	red clover	corn	oats
Field 3	oats	red clover	corn

On rolling or steep slopes a five-year rotation would look like this, breaking up the fields into five equal groups instead of three, and using alfalfa instead of red clover:

	1st Year	2nd Year	3rd Year	4th Year	5th Year
Field 1	corn	oats	alfalfa	alfalfa	alfalfa
Field 2	oats	alfalfa	alfalfa	alfalfa	corn
Field 3	alfalfa	alfalfa	alfalfa	corn	oats
Field 4	alfalfa	alfalfa	corn	oats	alfalfa
Field 5	alfalfa	corn	oats	alfalfa	alfalfa

Notice that in each field, corn follows the legume sod crop, and oats, which does not need plowing, follows the corn. Since the oats also acts as a "nurse crop" for the clover or alfalfa, which is sown at the same time, the forage crop follows the oats.

Since red clover normally lives only two years, alfalfa is used in the five-year rotation instead of clover. Whatever crops are chosen will be picked for their suitability to the soil, climate and topography, but they will usually include a legume, to keep up the nitrogen level.

The observations under the topic of plowing, regarding working land which is too wet, also apply to cultivation, though to a somewhat less extent, because of shallower digging. The old theory used to be that cultivation was both to destroy weeds and to maintain soil moisture, but controlled tests at several U.S. agricultural experiment stations have settled the argument. It is now well established that the only purpose of cultivation is to keep down the weeds. Experiments have shown that cultivation merely to stir up the soil and preserve a "dust mantle" does not actually make more moisture available, and hence is unnecessary. In fact, excessive cultivation is harmful, because it tears the life-giving roots, and brings more weed seeds near the surface, where they can germinate.

In a state of nature, plants compete for the available sunshine, moisture and nutrients. Exactly the same fight for survival occurs in a cultivated field, hence the farmer strives to give his crops the advantage over the weeds. Oats and other small grains are planted as early in the spring as possible, which gives them a head start on weeds which do not germinate until warmer weather. By the time the weeds germinate, they will be smothered by the oats. Since corn is planted about the same time the weeds are beginning to flourish, it must be cultivated, or the weeds will steal all the moisture and nourishment, and the corn crop will be cut to a quarter or less.

Strongly dominant weeds like Canada thistle can be killed by chemical sprays such as 2,4-D, which will not hurt the grain seeding. Chemicals are now also used to some extent instead of corn cultivation, the spray being directed below the corn leaves.

Erosion control has been harped on so strenuously in recent years that it hardly seems necessary to mention it, but the beginning farmer should know the signs and the remedies. The three chief kinds of erosion are wind, gully and sheet erosion. Cultivated light soils in windy areas can be "tied down" by a cover crop such as rye for fall, winter and early spring. Shelter belts of trees can also be planted. Each one foot of tree height will protect thirty feet of ground.

Gully erosion, where the land is split by jagged valleys, is all too easy to recognize, and not too quick to cure. Erosion control dams, horizontal terracing of the land to make the water "walk" instead of run to grassed sluiceways, strip cropping and contour plowing are the usual methods. U.S. soil conservation districts will help with plans and financing to control this most serious form of erosion.

Sheet erosion is much more difficult to detect, but can occur on any field with a slight slope. Instead of gullies, a thin layer of precious topsoil can be washed away by heavy rains. The only answer is to keep the land in grass as much as possible, instead of in cultivated row crops. (See Chap. 16, "Grassland Farming Is Easier.")

Fertilization and tilling are the basic elements in maintaining soil. There are, in addition, some minor aids, such as angleworms, mulch tilling, and the like, which contribute their bit, but should not be allowed to become obsessions.

ADDITIONAL READING MATERIAL: (free from U.S. Dept. of Agriculture) *Summer Crops for Green Manure and Soil Improvement*, F 1750; (from

Supt. of Documents) *Laying Out Fields for Tractor Plowing*, 10¢, Cat. No. A 1.9:1920; *Plowing With Moldboard Plows*, 10¢, Cat. No. A 1.9:1690; *Cover Crops for Soil Conservation*, 5¢, FB 1738; *Terracing for Soil and Water Conservation*, 10¢, FB 1789; *Ways to Till the Soil*, 5¢, YS 1956.

-16-

GRASSLAND FARMING IS EASIER

A method which saves your back and the soil at the same time.

Good hay and pasture take much of the place of expensive grains and concentrates.

A greatly simplified, prosperous and permanent type of agriculture.

Easier work for the farmer and the hired man, and a great saving in storage space required.

The problem of barn driers and hay finishers.

Tripling the productiveness of pastures.

Let the cow harvest the crop.

Recommended seedings for spring, summer and fall.

Hay cut too late and weathered too much makes poor feed.

New devices which checkmate the weather and ease the job of haying.

"Pickling a pasture" in the silo.

Grassland farming controls weeds more easily.

THE WHEEL has come full circle. Our nomad ancestors, who pastured flocks and herds wherever there was grass, have descendants who are now worshipers of grassland farming. It's a good farm religion. Good, that is, for America, where cloudbursts and heavy snowfall make erosion a more serious problem than in areas where rainfall is gentle and snow scarce. Grassland farming, which involves "processing" fodder through animals and then marketing the animals or their products, is not the key to the world's food problem, because it takes many fewer acres of grain, used directly as human food, to support humanity than when crops are run through an animal to make more complex products. Nevertheless, because of conditions peculiar to large sections of America, grassland farming is highly desirable here.

In its essentials, grassland farming is merely a refinement of the doctrine that "farming is a way of life." Grassland farming seeks to promote soil conservation, easier agriculture, and the same income as the more strenuous forms of tilling the soil. Basically, grassland farming means using level fields for row and grain crops in rotation, leaving the rolling lands in grass much of the time, keeping hillsides likely to wash in grass all the time, and reserving the steepest slopes for trees, with cattle fenced out. Most of the material in this chapter is derived from the University of Wisconsin agricultural bulletin on grassland farming.

While it is true that the University of Wisconsin College of Agriculture had in mind especially the rolling hills of southwestern and western Wisconsin, the principle can be applied to any rolling terrain, and is entirely applicable also to flatlands. For rolling lands, grassland farming is just plain common sense. It is the only method of preserving the precious topsoil, and it enriches the soil while it eases the tasks of the farmer. While not so vital to level fields, it remains a method of farming which conserves soil and human energy at the same time, without reducing farm income.

Basic to grassland farming is the fact that plenty of good hay and good pasture can take the place of most of the grains and the expensive protein concentrates ordinarily used in farm operations. It involves sound and wholesome cropping and soil conservation practices, which stop erosion caused by leaching and excessive plowing, while it not only maintains the fertility of the soil, but increases and improves it.

Grassland farming, in fact, could be described as a greatly simplified, prosperous and permanent type of agriculture. Because it involves much use of machinery and elimination of drudgery, it is the type of agriculture that should attract and keep youths on the farm.

Instead of leaning heavily on corn and oats for his feeding ration, the grassland farmer emphasizes hay and pasture—in fact, he expects to get fully 80 per cent of his feed requirements from hay and pasture. This means more land in grasses more of the time, and some land in grass all of the time—grasses being defined as the true grass-legume mixtures, as well as pure legumes like clover and alfalfa. The grassland farmer plants more acres to high-quality hay, and he betters his pastures by proved methods, to produce from two to three times the amount of forage ordinarily obtained from them.

One of the great beauties of grassland farming is that the agriculturalist is devoting himself to basically good farming practice. He uses longer crop rotations, and practices manuring, liming, fertilizing, strip cropping, and contour plowing. He uses cultivation and drainage only where necessary. Instead of considering pasture and forage as something reserved for the marginal or poor lands, he recognizes that the best land is none too good for hay and pasture.

When he plants corn or small grains, the crops will be more productive because of the enrichment of the land by preceding crops of good forages. He cuts his feed costs because his home-grown forages give the cheapest and most economical feed for all classes of livestock. If bad times come, and he needs a financial "shot in the arm" like a big corn crop, he can draw on the "soil bank" which he has created by steady use of legumes.

Nor has the grassland farmer sacrificed flexibility with his program. His forage crops can be used for hay, pasture, grass silage, seed production, or merely soil improvement, depending on his needs.

Grassland farming, of course, does not mean the complete elimination of grains. Oats and similar small grains will continue to be used as companion or "nurse" crops in establishing stands of grasses. When corn is planted on level lands, it will be a heavier producer, thanks to the "soil bank" effects of several years of legume growing, which stores nitrogen in the soil. In general, however, some corn acreage will be shifted to production of grasses and small grains

which require less cultivation, hence less baring of the soil to wash-
ing rains and eroding winds. In the same way, hilly portions of
present permanent pasture will be retired into permanent woodlots.

What does all this do to total livestock numbers on a particular
farm? Possibly nothing at all. Instead of depending on the feed bin
or silage from the silo, dairy and beef cattle will spend more of their
time on good pasture, and will learn to eat grass silage instead of
corn silage. Pigs will be out on pasture, getting a good portion of
their protein requirements, rather than gorging at the self-feeders.
Even poultry can thrive on pasture, because of the high protein
content of immature forage. In fact, because of lower feed costs
and increased crop production possible with grassland farming, the
same farm may be able to support more livestock than before, which
means more income for the farmer.

One of the decisive factors in favor of grassland farming is easier
work for the farmer and the hired man. The farmer who turns to
grass, if he invests in the right machinery, can look forward to doing
only half as much haying work as his forebears, even while he is
producing bigger crops. Forage harvesters and hay balers have taken
the heavy work out of the former tasks of hay loading and mowing.
Studies at the universities of Ohio, Iowa and Wisconsin indicate that
even without a forage harvester or hay baler, a farmer with a
buck rake, mower and electric hay hoist can cut his labor in half
over the former method of harvesting long hay with a wagon and
hay loader.

In addition to the labor saving, there is a tremendous saving in
storage space, thanks to modern methods of hay making. In other
words, it is possible to store from two to three times as much hay
in the same-size barn as was possible under the old loose hay system.
A ton of loose hay fills 450 cubic feet, while a ton of chopped hay,
produced by a forage harvester, occupies only 250 feet, and a ton
of baled hay fills just about 123 cubic feet. (The space occupied
by baled hay is not always necessarily so compact, because of the
bad habit of custom balers of making the bales loose instead of
tight. Sometimes a loose bale is necessary because of high moisture

content in the hay, but often it merely reflects the desire of the custom baler to charge for more bales per acre.)

While barn capacity can be doubled or nearly tripled by putting up chopped or baled hay instead of loose hay, take a good look at the beams and flooring before you pile on this added weight. Many an ancient barn floor, built to stand a full load of loose hay, has collapsed when the owner piled on twice the weight in baled or chopped hay.

Even with a power elevator from ground to haymow, baled hay still requires considerable heavy work when it is stowed away in the barn, but the work is less than with loose hay, and the baled hay is very easy to feed out. Anyone who has wrestled with a pitchfork, trying to dislodge enough hay from the mow to feed the cows, will appreciate the ease of dropping a few bales down the hay chute, ripping them open, and distributing them to the livestock.

Chopped hay, blown from the forage harvester into big self-unloading box wagons, and from the wagons, by another power blower, into the barn, takes no hard work in filling the haymow, but needs a strong back at the scoop shovel in feeding it out. The chopped hay, especially when cut into sizes three inches or longer, has the advantage of being consumed entirely by the livestock, stems and all, while the same animals, fed loose hay, are likely to pick and choose, spoiling what they do not eat. A disadvantage of chopped hay is that it is usually dusty.

Forage harvesters offer the greatest flexibility, and perhaps the least exertion—if you don't mind riding a tractor seat. A chopper with suitable attachments can be used for corn silage, dry hay, grass silage, and straw. The big wagon boxes trailing the harvester can be made with slanting floors, like a coal gondola car, so that the chopped hay spills out by gravity into the blower which whisks it into the barn.

A neighbor of mine, who had always made loose hay, with his wife driving the tractor while he loaded atop the hayrack, acquired a forage harvester recently, and was able to do all the work himself. He commented to me:

"I never made hay so easy before in all my life."

Just as barn floors and walls must be extra strong to handle baled or chopped hay, so must silos sometimes be reinforced to withstand the sideward pressures of grass silage.

Should you go in for "barn driers" or hay finishers? The idea is appealing, because it is possible to cut hay in the morning, wilt it in the noonday sun, put it in the barn in late afternoon, and start the forced air currents circulating through the new hay to remove the excess moisture which would otherwise result in spontaneous combustion or spoiled hay. The vogue of barn driers—air ducts in the barn floor putting a forced draft through the newly stored hay —began in the Tennessee Valley, where they were popular and effective. Nevertheless, the transfer of barn-drying methods from the comparatively low barns of Tennessee to the higher-peaked barns of more northern regions has been attended by considerable difficulty. A system good for fifteen feet of hay didn't always work when applied to twenty or thirty feet. Farmers who installed hay driers in big barns have sometimes been forced to resort to supplementary ducts, to insure proper drying.

When a new barn is built, drying ducts can be installed without too much expense, but to convert an old barn to a drying system may mean costs way out of line with results. Before installing such a system, it would be well to see whether any neighbors with the same-size barns have installed driers—and with what luck—and it would also be a good idea to get the advice of the county agent.

A new development which holds much promise is the specially-built hay drier. Fashioned like a silo, it has a central hollow core for forcing air through the partly cured hay. It can be filled with a blower and emptied by gravity. The hay drier has several advantages over grass silage: it gets away from the messy leakage around the silo, typical of grass silage; it almost completely checkmates the weather; and it produces hay of high protein and bright green quality, which means more profits in the milk pail and in fattened steers.

Along with hay production, the grassland farmer needs to pay attention to his pastures. The trouble with the ordinary pasture, composed usually of blue grasses and a little red top, is that it makes perhaps three-quarters of its total seasonal growth before midsummer. After that, the cows have lean pickings. Such pastures do not afford true grassland farming—they are just an insult to the innards of the cow, and mean supplementary feeding through the rest of the hot summer months, either by cutting green corn, feeding hay which the farmer has already put up, or emergency pasture such as Sudan grass.

The so-called "permanent pasture" on many farms is often a sparsely covered hillside, a neck of woods, and a little bottom land too wet for the plow. It may be only partly cleared of stumps and rocks, or be overgrown with coarse weeds. The farmer who depends on these for all-summer pasture is likely to find his herd critically short of feed during the hot, dry days, and produce severe over-grazing.

Although many farmers hesitate to use their good cropland for pasture, nevertheless good pasture ranks highly with other crops in yield of feed per acre. For instance, the dairy farmer, interested ultimately in how much milk he can get for an acre of crops, should know that if a heavy cow averaging 20 pounds of milk daily can get 150 days of pasture from an acre, this acre has produced in nutrients the equivalent of 48 bushels of ear corn, or 95 bushels of oats.

And the yield that the farmer is interested in has been produced at lower cost, because the land does not have to be plowed, prepared and seeded each year, no cultivation is required, and the cow harvests the crop! Furthermore, if planted to corn or oats, this land would not afford a balanced ration, but would require additional protein, whereas the pasture itself supplies adequate protein.

The economy and nutrient value of good pasture applies equally to good hay. Consider that each acre should produce two tons of good alfalfa hay, which will equal 150 days of grazing per cow, or 95 bushels of oats or 48 bushels of corn, and will in addition have much more protein than the corn or oats.

Alternating the good fields with pasture and hay is the easy way of grassland farming, but what of the marginal fields, the hillsides and other places unsuited to the plow? These also can be made to be much more productive. Under proper "renovation," thin, unproductive sods on hilly land can be made to produce from two to three times as much forage. The method in renovating is to establish, without plowing, a new seedbed based on superior forages like alfalfa, sweetclover, medium red clover, Ladino clover, Lespedeza, smooth brome grass and timothy.

If the terrain is open pasture land, the sod can be torn up by thorough going over with a disk and spring-tooth field cultivator. Rough and hummocky pastures, and those filled with brush, rocks or stumps, will need an A-drag or a bulldozer or both, to clear them. The thorough cultivation is needed to destroy most of the grassy vegetation and to work in the lime, phosphate and potash which such soils require in order to promote a successful stand of legumes. By the time you are through with disk and spring-tooth, the seedbed should look as if it had been plowed and harrowed, except for the loose sods on the surface. If the pasture is run-down, thin grass sod, it probably can stand two or three tons of lime, and 300 to 400 pounds of commercial fertilizer per acre. Test the soil to see just what mixture of fertilizer is needed. (See Chap. 15, "The Living Soil.")

For thin, poor soils, seed a mixture of twelve to fifteen pounds of the common biennial sweetclover, five pound of medium red clover and three to four pounds of timothy. Poorly drained soils, not suited to alfalfa, can use a mixture of four pounds of medium red clover, one to two pounds of Ladino clover, and six to eight pounds of brome grass or four pounds of timothy.

For well-drained soils which have been properly fertilized, you can use eight to ten pounds of hardy alfalfa, six to eight of smooth brome grass and five pounds of medium red clover per acre. If there is plenty of moisture, an added pound of Ladino clover per acre will improve the yield and quality. Wet soils will grow reed or canary grass for pasture. If a cultipacker seeder is used, you can reduce the seeding rate about 30 per cent.

Note: The legumes like clover and alfalfa are never seeded alone, because they will not produce good stands if seeded by themselves. They are always seeded in combination with grasses, such as timothy or brome grass. Likewise, the legumes, which have the ability to obtain nitrogen from the air and store it in the soil in nodules along the roots, will not do so unless the seed is inoculated ahead of time with a nitrogen-fixing compound bacteria suited to that strain. (The inoculation process is simple.)

While you are seeding these grasses and legumes, seed also a bushel and a half to two bushels of oats per acre as a nurse crop, which will help to keep down the weeds, check erosion, and which can be harvested as grain or pastured. The nurse crop protects the seedling hay crop, and is harvested before the seedlings are more than a few inches high.

Remember that in converting farm crops and pastures into meat or into milk, you are interested in the "total digestible nutrients" per acre, for that is what will make the meat or the milk. From pasture, these nutrients cost less than half as much as they do in the form of hay and silage, and less than a fourth as much as in grains or concentrates. That is why the wise farmer plans his farm management so that he will have continuous high-quality grazing for as long as possible.

Local conditions will of course have a lot to do with it, but here are sample suggestions of what can be provided, in addition to the permanent pasture:

For early spring grazing you can use common rye sown in the fall, or for a week, when the ground is firm, pasture your hayland. This should be done only if the stand is vigorous, but it can be grazed heavily, and will delay the time of cutting, which will produce finer and better-quality hay.

In spring and early summer you can use the ordinary mixtures of grasses and legumes commonly sown for rotation hayland and pasture. In addition, there is the possibility of spring-sown small grains, sweetclover, Ladino clover, and reed canary grass.

In summer and for summer and early fall grazing, the possibilities include Sudan grass, Balboa rye, sweetclover, pasture and hayland mixtures of alfalfa, Ladino clover, and smooth brome grass, or reed canary grass, or mixtures of soybeans and Sudan grass, or soybeans, Sudan grass and corn. Other choices are dwarf Essex rape or millet.

Fall grazing could be the hayfields or the rotation pastures which you plan to plow the next spring, common rye, reed canary grass, dwarf Essex rape, and later, after frost or cold weather has retarded growth, the vigorous growths of new seedings and established stands.

In grassland farming you will keep your livestock on pasture just as long as possible, but you will also be putting up a lot of hay which, because of uncertain weather conditions, is usually the most variable in quality of any crop on the farm. On many farms the greatest single loss year after year is the nutrition loss that comes about in the curing and handling of hay—"the feed that is grown but never fed."

Shattered leaves which fall off before the crop gets into the barn mean that much less protein, minerals and protective vitamins. Between 60 and 75 per cent of the protein in hay is found in the leaves rather than the stems. A bright green color and plenty of leaves mean high-quality hay.

If the hay is cut past the proper time, it has already lost much of its protein, and hence its feeding value. Sun bleaching, spoilage and leaching by rain and heavy dew, are added to the losses caused by shattered leaves. Poor-quality hay can cause various diseases in calves, lambs and colts, and difficult breeding and general unthrifty conditions in mature animals.

What can you do to insure getting the best-quality hay in your barn?

Several new devices help to shorten the time the hay is exposed to weathering, and also reduce the amount of labor required. One of these is the combination mower and stem crusher, which passes the cut hay between two rollers, crushing the stems so that they dry almost as fast as the leaves. Experiments indicate that the time of hay curing is reduced by the crusher by as much as two or three hours. Forage harvesters make for more uniform curing, by chopping up the stems. Mow curing by forced drafts, and the new hay driers, permit cutting and storing the hay in one day.

One great danger of the new devices: there is a temptation to wait past the proper cutting time, until a custom baler or forage harvester is available. Such practices can lose as much as half the feed value of the hay. A further error is to cut a whole field at once, which is handier for the baler, but results in a big loss in quality for the last of the hay to be baled.

Another method of cheating the weather is by using grass for silage. Some farmers, conscious of the frequent rainy spells of June, put the first cutting of hay into the silo, and fill their barns with the second cutting, when there is less danger of rain. Since putting forage crops into the silo eliminates the need of field curing, losses of leaves and leaching and bleaching damage are cut to a minimum.

Other advantages are that storage space is only a third that of hay, there is virtually no fire hazard, the feed is succulent, and keeps its feed value during long storage better than hay. Crops full of weeds also make better silage than they do hay. For the grassland farmer, putting some of the hay into the silo green permits him to harvest excess pasture not needed at the moment, at a time when it is in best condition.

Nevertheless, grass for silage has certain disadvantages as well. Unless the silo is strong, it will need additional reinforcing to withstand the extra pressures of grass silage. There is somewhat less palatability in grass than in corn silage, money must usually be spent for added preservatives, and it is difficult to control the moisture content.

Because forage legumes and grasses have a lower sugar content, and lower total digestible nutrients—although they have higher protein and carotene than corn silage—they do not assist the lactic acid organisms as well as corn silage does. This can be somewhat overcome by partial wilting of the crop, to cut the moisture content to 60 or 70 per cent before putting in the silo. Other methods are additions of molasses, phosphoric acid, ground corn, corn and cob meal, barley or whey while ensiling.

One additional benefit of grassland farming deserves some mention, and that is the effect in controlling weeds. If the fields are properly fertilized and managed, the grasses should be able to control most weeds by competition. For those which are more persistent, like thistles, there are now chemical weed killers like 2,4-D, which control the weeds without injuring the grass or livestock. Mowing is an old-fashioned way of controlling weeds, but it should not be considered, in the light of present methods, except for cutting Canada thistles in good stands of alfalfa.

A cheerful thought is that quack grass, very bad in a cultivated field, is much sought after in its young stages by cows for forage. Cows will ignore good alfalfa, which has a somewhat bitter taste while growing, to search out and crop any sprigs of quack grass in the field. In fact, they will keep the quack grass cropped down to the ground.

ADDITIONAL READING MATERIAL: *Pleasant Valley* and *Malabar Farm,* by Louis Bromfield.

-17-

THE HUM OF THE HIVE

"Let-alone" beekeeping is the easiest way.

Buy new swarms and hives, to avoid disease.

Keep flight lines out of the farmyard.

A pan of sugarwater near the hive will keep the bees going until the flowers are ready.

Once a week is enough to inspect the hives, under the "let-alone" system, but wear clean clothes, and be clean yourself, for bees are sensitive to odors.

You don't need fancy equipment to care for the bees or to extract and preserve the honey.

When the bees swarm, be ready with a spare hive.

A good hive will produce between 50 and 75 pounds of honey during the summer.

Cut a winter flight hole near the top of the hive.

Besides producing honey, bees are essential for fruit and legume seed production.

YOU may not want to become a full-time beekeeper, but don't be afraid to experiment with a few hives. With almost no technical knowledge you can follow the "let-alone" plan of raising bees, and do pretty well at it. Under this system you set up the hives in a desirable location, and pay attention to them only when you feel like it. Of course you won't get the production that the professional squeezes out of the colony, but you won't do much work either.

No matter how cheap the price, don't buy an established hive of bees from your neighbor or anyone else. The chance of importing disease or run-down equipment is too great to be worth the risk. The mail order stores sell knocked-down hives which you can assemble, along with frames and sheets of wax to fill them. The bees, all packaged, will come by mail. Since bees are now bred for good nature as well as production, specify "gentle Italians" when you order them. It may save you many a sting.

Get the hive materials early in the spring and assemble them. Any number from one to four will do for a start, but get one more hive than you plan to order packages of bees, in order to have an extra hive available in case of swarming. The mail order houses also have good, inexpensive pamphlets on care of bees. A good hive location has sun, with some shade during the middle of the day. The hives should be away from chickens, which eat bees, and, of course, not in a place where cows, sheep or pigs can knock them over or disturb them. If possible, have the ground slope toward the hives, so that the bees will be "sailing downhill" as they approach, loaded. Sometimes, with an especially heavy run of nectar, bees load themselves so heavily that they can barely make the entrance platform, which should be low enough so that stragglers can crawl up to it if they can't quite make the hive. Try also to place the hives so that trees, buildings or other barriers will keep the flight lines from crossing things like the clothes-drying yard or the center of the farmyard.

The mail order house will tell you the correct time of year in your locality for starting the bees. The "package" bees are raised in the South, and are shipped North in the spring, with each package containing a queen and between 2,000 and 3,000 workers.

The package bees have directions for releasing the queen, who is in a separate package, and the worker bees. The total of bees in the package is not enough to produce any amount of honey, and for the first six weeks or so they will be building up the number of workers in the hive. Unless flowers of plants or trees are available, the bees at first may need artificial feeding, which you can furnish by setting a pan of sugar water in front of the hive. By midspring the bees should have built up a colony of 15,000 to 20,000 workers, and be in full production.

The bottom box with which you start, the hive proper, is the place where the egg-laying and raising of new workers takes place. When the colony is built up, the bees are ready for storing honey, done in upper boxes, called "supers." These are slightly shallower than the hive, and are filled with standard frames for honeycomb. The first thing the bees will do is seal the edges of the layers of supers.

Under the "let-alone" system of beekeeping, you take a look at your hives once a week or so. If you see a lot of bees coming in and out of the entrance, put on three or four supers, and let the bees do the rest. By mid-July you can start using your own honey. Take off the bottom super, remove the frames, slice the caps of the combs off with a knife heated in hot water, and let drain through a strainer into whatever jars are available. Each hive, during the summer, should produce between fifty and seventy-five pounds of honey, but you must leave about twenty-five pounds for the bees to feed on during the winter.

Even the gentlest bees are very sensitive to colors and especially to odors. And they don't like cloudy or rainy weather. When several days of bright weather, a good honey-flow time, are followed by clouds and rain, this is no time to monkey with the bees, for they will be as mad as hornets, because frustrated. The time to change supers, or do anything else about the hives, is around noon on a sunny, rather windless day. Most of the bees will be out gathering nectar and those at home will be sociable and in good humor.

Since bees are so sensitive to odors, you can save yourself a lot of trouble by being reasonably clean, and wearing fresh, clean clothing. You don't need to scrub up like a doctor for an operation, but simple cleanliness pays off in working around bees. Human sweat seems to be their pet aversion, so walk, don't run, to the apiary.

The only indispensable tool is the special metal pry, like a screwdriver, for loosening the supers which the bees have glued together. If you can move without too much banging, you may be able to do all your work without any special protection. While you are learning, however, you would probably be wise to wear gauntlet gloves, a shirt that buttons tight around the neck, and a hat with a cheesecloth veil which you can gather round your neck. Such equipment is on sale, but you can easily make it yourself.

Special smokers are available, for blowing smoke into the hive to quiet the bees, but you can achieve almost the same results by burning a few rags near the entrance, so that some of the smoke is wafted inside. Many beekeepers don't even bother with a smoker. They wear clean clothes, move quietly, and the bees are hardly aware that the hive is being disturbed.

Since beekeeping goes back beyond recorded history, there is plenty of printed material to help or inspire you, but nothing can beat your own personal observation, as you learn their habits. Even though the worker's span of life is only about six weeks, she crowds a lot of excitement into it. The bee that discovers a new field of nectar comes back to the hive and does a special "honey dance" on the front doorstep, to tell the other bees about it.

One of the greatest thrills in beekeeping is the hiving of a swarm, which is about the only trouble the beginner has to worry about, anyway. One way to prevent swarming is to pile on the supers, so that the bees have such a big backlog of comb to fill that they don't bother to raise a rival queen bee. However, you might as well be prepared for the fact that some of your hives are going to swarm, and keep an extra hive in reserve for this great event. Often the bees will leave the hive, and cluster immediately on the nearest branch. If this happens, set up your new hive, cut off the branch, and either place it in front of the new hive, or gently shake it into the hive, depending on how courageous you are about bees.

Some beekeepers cut off the branch, dig into the crawling mass of bees until they find the queen, and start her up a cloth incline toward the new hive entrance. The workers, as soon as they hear the special sound of the queen bee going into the hive, will march up after her with a mighty hum that is one of the most impressive sounds of beekeeping.

Of course, the bees may, by misfortune, not light on the nearest branch, and may take off across the countryside in a pell-mell howling race. (In the old days on the farm it was the job of the farmer's wife to watch for swarms, and she and the brood of children would dash out banging on dishpans to drown out the leading noise of the queen bee. The theory was that if they could make enough noise, the workers would be unable to hear the queen, and would settle close by.)

If the swarm should escape, you may be able to follow it by car (the bees move in a dense, zigzagging dark cloud, skimming the treetops). Take the new hive with you, and when the bees cluster on a branch, shake them into the hive, cork it up, bring it back, and unplug the opening after nightfall.

After you have taken the surplus honey in the fall, and reduced the size of the hives to the original brood box and one super, you can prepare the hives for winter. You can pack leaves or straw around the windward sides of the hive, but if the climate is not too cold, and the hives are not in an extremely exposed condition, this precaution is not necessary. However, the bees like to take short exercise flights on sunny days throughout the winter, and a good precaution is to cut a flight hole slot near the top of the hive, to permit them to get out freely. The normal entrance at the bottom cannot be depended on in winter, because it often gets clogged shut with dead bees. In early spring check the colony to see whether the bees survived the winter. If they didn't, or if they look enfeebled, order a fresh package, which costs little and will give you a good vigorous start on a new colony.

Professional beekeepers have a lot of fancy and expensive equipment for extracting the honey, but you can get along without it if you want to. A hot knife will slice the caps off the frame combs, and you can set them to drain, which will bring out most of the honey. As a special treat you may wish to try one super of "box" combs, which make an attractive addition to the table or for sale.

Besides having honey for sale, it can be a solid addition to your food supply. Children soon learn to love it on cereal and griddle cakes, and it can be used instead of sugar in cakes and cookies, and in much other cooking.

By watching to see what flowers the bees are visiting you can gather different kinds of honey, each with its own flavor. The big honey flow is usually produced by blossoms of sweet or red clover, but some people prefer the delicate flavor of basswood, and there are a few who enjoy the dark, full body from tulip trees or a field of buckwheat. Put on a fresh super for the flavor you want, and take it off when the honey flow from that flower stops. When you get fond of honey, you will find yourself using as much as ten to twenty gallons a year for a family of five.

In addition to supplying honey, bees play a vital role in farming as pollinators. They are essential in spring for fertilizing fruit trees, and in the summer, if you are raising a field of clover or alfalfa for seed, you won't get any merchantable quantity unless there are plenty of bees around to pollinate the flowers. Some obliging commercial beekeepers check with their neighbors, and if a farmer is planning to keep a field of legumes for seed, will move up several colonies, to make sure of a seed crop. And some beekeepers rent out swarms for this purpose, instead of merely lending them.

ADDITIONAL READING MATERIAL: (from Supt. of Documents) *Productive Management of Honeybee Colonies in the Northern States*, 10¢, Cat. No. A 1.4/2:702.

-18-

ORCHARD, WOODLOT AND FENCEROWS

ORCHARD

*In starting an orchard, "whips" will mature as quickly and cost less
than older trees. Pick varieties good fifteen years from now—
if you can.*

Raise cash crops between the rows.

If there are not enough bees for pollination, raise your own.

You can sharecrop your orchard to an expert.

A few trees for a north slope or an unused corner.

THE FARM WOODLOT

Keep the cows out if you want a woods.

Thin out leaners and weaklings for fence posts and firewood.

Look out for "widow makers" when felling trees.

*Season firewood a year before burning or selling, to get the most
heat and the most money.*

FENCEROWS

*A good fence is worth the trouble, but does not need to be "horse
high, pig tight and bull strong."*

*You can dig a fence post hole with a quart of water, but spring is
the easiest time to build fence.*

*Use locust wood, chemical treatment, or concrete posts, or you will
be rebuilding your fence every dozen years.*

An electric fence is cheapest and easiest to build.

Don't overlook the pleasures of a wooded fencerow.

ORCHARD

IF YOU want to be a full-scale orchardist, getting most of your
living from the fruit of trees, the quickest way is to buy a mature
orchard, but it may be the most costly. The other way is to start
from scratch with "whips" or small trees. It is slower, but you learn
the business before it overwhelms you. Still another possibility to
be considered here is the small fruit orchard as a supplement to
other farm operations, or as part of a subsistence homestead. Most
of the observations will apply to all three methods.

The dangers to the novice in buying a big orchard in full swing
are so obvious that they scarcely need to be touched on. Human
nature being rather crafty, you may be buying a dead horse, as
regards either the trees or the potential market for fruit. Consider
also that you must begin immediately with about eight required
sprays a year, the business of pruning, and especially, the marketing,
all of which takes a certain knack. Unless you have a lot of self-
confidence, think twice before you do it.

Starting from scratch with small trees gives you two main jobs
to do: caring for the trees so that they will produce early and
heavily, and trying to make a living until you have quantities of
fruit to sell. Unlike almost all other branches of general farming,
climate is all-important for the orchardist. The good fruit-growing
areas are near, and preferably to the east, of large bodies of water,
so as to insure a mild climate. An easy way to find out whether the
climate is suitable is to check on whether there are near-by productive orchards. If there are none, you are likely to be in the wrong
place for fruit trees as a main crop.

Old-line orchardists insist that "whips"—the one-year-old stocks
of trees—will produce just about as fast, and at much less original

cost, as the more expensive four- and five-year-old trees. They certainly don't look like much, planted twenty-four, or preferably thirty, feet apart in rows, but they have a small root system, and hence are less likely to be set back in transplanting, and after they once get started, will equal more mature trees. Incidentally, they can be protected from rabbits and other gnawing pests in winter by wrapping a heavy magazine, like the *Saturday Evening Post*, around the trunk, which is cheaper than wire netting—if you happen to have a surplus of magazines.

The full-scale orchardist has a real problem in choosing the varieties of fruit to put in. His difficulty is that he must gauge the public taste fifteen years from now, rather than depend on what is immediately popular in the market. Kinds of apples that are in favor now, for instance, may be lightly regarded by the time his trees are bearing. Before investing in any trees, the man planning a full-scale orchard should certainly consult his county agricultural agent and the horticultural department of the nearest agricultural college to get suggestions on what to set out. Part of the consideration, also, will be varieties suitable for that area and climate.

Whatever combination of trees is selected, the orchardist should make sure of sufficient pollinators. Some varieties are hard to pollinate, and thus should include, usually as every fourth and fifth row, two rows of good pollinating trees, to insure fertilization of the blossoms. Government pamphlets have lists of the good and the difficult pollinators, and which ones combine well.

Since practically all fruit trees, with the exception of some wind-fertilized plums, are fertilized or pollinated by bees, the fruit grower must be near bees, or else keep his own hives to make sure of loaded limbs. The two occupations, incidentally, go well together. (See Chap. 17, "The Hum of the Hive.")

There is no space here to go into the many details involved in full-scale fruit farming, but there are plenty of pamphlets and books on the subject, to which the reader is referred for further information.

What of the farmer, full-time, part-time, or just subsistence, who wants to use a north slope too steep for grazing or crops, or has a corner near the house which could be turned to fruit trees? If you get more than a dozen or so small trees, it means acquiring a full spraying outfit, or hooking up with a co-operative spray ring, or a commercial sprayer. However, even here it is possible to be in the fruit business without touching a tree or an apple. I know of individuals with established orchards who "farm them out" to retired orchardists, who care for the trees, gather the fruit, and market it, on a percentage basis. Such an arrangement would require several dozen trees to make it profitable, but it is one way of making a neglected slope bring in some income, without causing the owner too much work.

In addition to the fruit crop, sheep can also be raised in the orchard. Pick a short-legged breed such as Shropshires, or they will make lacework of the lower branches. The sheep will keep the grass down under the trees, contribute fertilizer, clean up the windfalls, and bring an extra dividend in wool and mutton.

For the farmer planning a small orchard as a side line or to farm out to an expert orchardist, the same thorough grounding in fundamentals, through more extensive books and pamphlets, is recommended.

However, even if you do not plan any very extensive fruit growing, every farm should have half a dozen or so fruit trees, to fill in an odd corner or steep slope, to give color and a lift in spring, and to fill the fruit cellar for the fall and winter. Yellow transparents and the like for early summer pies and apple sauce. Wealthies, Saxonians, Jonathans and McIntoshes for late fall eating and cooking, and "good keepers," such as Winesaps, Spitzenburgs and Northern Spies for the long winter months.

One possibility, available in recent years, is the dwarf fruit tree. These trees grow no higher than eight feet, but bear heavily and can be put in places where a higher tree would obstruct the view or take up too much space. Some grafted varieties contain three or four different kinds of apples on the same tree, so a grower with only four or five trees could get a dozen different kinds of apples.

Rabbits and mice are the two chief enemies of growing orchards, and the fall of the year is the best time to guard against them.

Protect the trunks of young trees with a roll of screening or an old magazine, to keep the rabbits from gnawing the bark. Branches likely to be reached in case of heavy snow can be painted with a mixture of resin and alcohol. For mice, which get into the roots and kill them, poison baits should be distributed early in the fall. Put eight or ten baits (poisoned wheat and the like) near each tree, in furrows and depressions, and cover with straw. The baits should be distributed in the early morning, since mice feed in the morning and early afternoon of bright and sunny days.

Your county agent has pamphlets on spraying for the pests common in your area. Pruning should be done as soon as you set out the young trees, cutting back almost half the wood, to bring the roots and top into balance. Unless you prune thus heavily, the disturbed root system will not be able to supply the branches with enough nourishment, and the tree will be sickly and backward. Cut close to the trunk, to leave no shoulder to rot later, "stagger" the branches so that they are not opposite each other and so that, seen from the top, they radiate like the spokes of a wheel. Besides cutting out the extra branches, cut back the branches which remain. A good heavy pruning when the trees are first set out should make it unnecessary to prune for the next four or five years. However, if you plant whips, do not trim back the whip or leader, since that is the growing portion of the tree.

The trees, regardless of the size of the orchard, will grow fastest if kept cultivated. Break up the sods in a circle about three feet in diameter around each tree, and as the tree matures, widen the circle. Crops like alfalfa or hay and small grains can be grown between the rows while the trees are maturing, or you can use the space for cash crops like strawberries and asparagus.

THE FARM WOODLOT

Like any other acres, the woodlot on the farm will produce much or little depending on the time you are willing and able to spend on it. Don't make the mistake of trying to turn the woodlot into a pasture. Livestock won't find enough nourishment there to pay you to fence it, but they will trample and kill all the young seedlings, and damage all the older trees that they can reach. A woods where cows run is on the road to death, for no replacements will grow under such conditions, and as soon as the stand of trees matures and dies, the woods will be gone.

The ideal woodlot is one which is constantly producing. Dead and dying trees can be cut down for firewood and fence posts. The living trees should be scrutinized to see which can come out, to make others better. Since all trees are in competition with each other for light and nourishment, take out the spindly ones whose crowns are below the general level, for they will never amount to much. In the same way, "leaners" should be cut down, to make more room for the straight-growing trees. Thin out, also, the clumps of small trees, and where four or five hardwoods sprout from one stump, cut down all but one or two of the best, leaving them to mature into good-sized trees.

Trees under 10 inches in diameter are good only for firewood, fence posts, pulpwood, or for making into excelsior. "Saw logs" are trees 10 inches or over in diameter, which can be sawed up into boards. If you have several such trees, you may be able to get enough lumber for a new hoghouse, for instance, out of your own woodlot. A tree 14 inches in diameter, with a "merchantable" height of 30 feet will produce 100 board feet of lumber—that is, 10 boards a foot wide, an inch thick, and 10 feet long, or their equivalent. Trees used for saw logs also furnish fuelwood in the branches and tops not used for boards. Roughly, there are about one to one and a half cords of fuelwood for every 1,000 board feet of saw logs. (A cord is a stock of wood 4 feet wide, 4 feet high and 8 feet long.)

The spindly, crooked trees which you take out to improve the stand can be made into fence posts (7-foot lengths) or fuelwood. Fence posts should be not less than 3 inches in the smallest diameter, if they are split logs, and if of oak, they should be 4 inches.

Since market requirements for wood vary widely in different parts of the country, consult your county agent's office, and the forestry division of the nearest agricultural college for advice on marketing excess timber which you do not need yourself. If you sell to some other person who will do the cutting and marketing, be sure you both understand what trees are to come out, and the values per tree. In general, if you sell "stumpage"—in other words, sell the standing trees without measuring them for diameter and approximate board feet—you will get the least, and the cutter may ruin the rest of the woods.

If your woodlot is too small for commercial cutting, or you just want to get out enough wood for fence posts or fireplace, certain precautions are still in order, to save not only the trees but yourself. Watch out, for instance, for dead branches which may fall and injure you (they are not called "widow makers" just for fun). When the tree starts to fall, step quickly to one side, and preferably behind another tree. Never stand directly behind and in line with the direction of fall, because occasionally the branches act as a spring, and "shoot" the butt back several feet when they hit the ground. Never swing an ax until you have removed all branches and twigs which might deflect the stroke and injure you.

The standard method of felling a tree is to make an "undercut" on the side of the tree toward which you want it to fall. The depth of this undercut should be about one-fourth the diameter of the tree. After the undercut is made, start cutting on the opposite side, about two inches higher than the undercut. The lean of the tree, weight of branches on one side or the other, and the direction of the wind, influence the fall. Try to avoid damaging smaller trees, or felling the tree so that it hangs in the branches of another. Even a tree which leans perceptibly can be felled in a direction opposite to the lean, if you make the undercut deep enough. This shifts the center of gravity, and the tree will start to straighten up when you make the full cut on the opposite side.

After the tree is down, saw or chop off the limbs, trimming the larger ones for firewood. Pile the small brushy branches in one spot, and leave them as a shelter for wild animals and birds. The former custom was to burn the brush, but this kills all growing things close to the fire, and may damage near-by trees. If piled closely, the brush will disintegrate in a year or two.

When the trunk is cleared of branches, saw it up into lengths you can handle (four feet for firewood, seven for fence posts), split the big logs with wedges and a maul, and stack it to dry. Try to cut next year's firewood this year, because green wood has only two-thirds the available heat that seasoned wood has. It takes from six months to a year for wood to season. If you are selling firewood, you should get a third more for seasoned wood, since it has that much more heating value. If you sell green wood, incidentally, you are likely to have some irate customers on your neck, when their fireplace logs sizzle instead of burn.

YOUR FENCEROWS

"Good fences make good neighbors," poet Robert Frost said, and he could have added that bad fences between neighbors make for bickering and bad feeling. If the fence is so weak that your neighbor's cows get through into your corn and alfalfa, and some of them bloat and die, he will rightly bear a grudge, even if he doesn't also sue you—and you may find more corn trampled than the value of a new fence. A poor fence just isn't worth the trouble it can cause.

Therefore your line fence, which separates you from your neighbor, should be stout and in constant good repair. The old definition of a good fence was "horse high, pig tight and bull strong," but this exceeds the legal requirements in most states. Usually a fence which is four and a half to five feet high, and which will turn anything but pigs (is anything but a concrete wall pig-proof, I sometimes wonder?) will fulfill the legal requirements. Your township clerk can furnish you the legal requirements in your state, if your neighbor is a stickler. Usually, line fences are jointly owned: that is, your neighbor builds and owns one-half, and you build and own the other. For uniformity and good neighborliness it is well to consult ahead of time, particularly as to the kind of animals likely to "run against the fence," to use the country expression. If you or your neighbor has horses, the fence must be extra-high and strong. Unless the pasture on their side is plentiful, horses have a penchant for leaning over any fence less than six feet high, and they can soon weaken the top wires. Their sole advantage is that they will not try to break through a fence, the way a cow, sheep or pig will do.

A good line fence has woven wire 4 feet high, with two strands of barbed wire (called "bob wire" in the country) 6 inches apart on top. Posts should be not more than 16 feet apart (preferably about 12), and at least 2 feet in the ground. Corner posts are thicker, have diagonal braces, and go down 3½ feet.

The best time to make fence is in the spring, when the ground is soft and digging is easy. If you have to build fence in the summer during a dry spell when the ground is hard, try this trick to make the digging easier: Ram a crowbar down a foot and a half into the ground at the location of each prospective post hole, and pour in about a quart of water. In an hour the ground will be moist enough to dig. If you have a long stretch of holes to dig, fill a milk can with water, haul it out to the site in the trailer, make your holes with the crowbar and fill them with water, and by the time you have done, the first hole will be ready to dig.

Use a fence stretcher (a block and tackle with special friction catches to grab the wire) to stretch the fence tight, or if there is room, you can pull it tight with your car or a tractor. Use plenty of staples, and drive them in tight, because cold weather contraction of the wires will pull everything loose otherwise.

Unless you plan to rebuild your fence every eight to ten years, it pays to buy treated posts or use a decay-resistant wood like locust. Ordinary posts will last ten to twelve years, while locust, and posts treated with creosote, will last fifteen to twenty years. The Portland Cement Company has pamphlets showing a simple rack for making concrete posts which will last a generation.

The interior fences on your farm, and those along the roadside, can be as strong or as weak as your fancy dictates, since any damages to your stock or your crops will be strictly on your own head. Three strands of barb, for instance, would be enough to keep cattle out of a woodlot, but four strands would be better to separate cattle from such delicacies as growing corn or clover. Incidentally, don't tempt cows to break down a fence by having corn so close to the other side that they can nearly reach it.

The simplest and easiest fence to build to contain livestock is an electric fence. This is usually of barb, a single strand about three feet high, and the posts can be as much as thirty feet apart, and only a foot and a half in the ground. The electric fence has the great merit of being easily put up or moved, and animals have a healthy respect for it. In fact, after it has been on for a couple of weeks, the current can usually be turned off, to be turned on again occasionally, if the animals get too careless with the wire. The disadvantages of the electric fence are that it may short and become useless, the current may fail, or a charging animal may break through it. Some odd specimens of livestock are occasionally found, too, which actually seem to relish the jolt of the fence.

The current is furnished by a control box costing from $10.00 to $20.00, which steps up the voltage of a three-cell dry battery costing about $3.00 and good for several months. The fence controller delivers a pulsating current which is on for only a fraction of a second, and repeats about every half second. The "hot" wires from the controller go to the barb strand, which is clipped to insulators on the posts. The ground wire of the controller goes to a pipe sunk a few feet into the ground, and the circuit is completed when an animal touches the barbed wire. Since weeds and other objects can short the system and wear out the battery quickly, the fence must be patrolled occasionally to see that there are no shorts. Some farmers use a power lawn mower to keep weeds from under the fence. All of the equipment, including fence controller, wire clips specially bent to fasten the barb to the insulators, and handles for electric gates can be purchased at any rural hardware store at small cost.

The stores also have meters for testing the potency of the electric fence without getting a shock, but you can "short circuit" this expense by merely holding a weed on the wire, and firmly planting your feet on the ground. The weed cuts down the jolt so that it is no more than a small tingle.

One fence controller will handle as much as two miles of wire, and if it is strategically placed, you can use it for several fields. It is better to educate the livestock to the electric fence for a day or two before turning them into a big field, or they may stampede and break through it. You can pen them up in a lane, for instance, where they come in frequent contact with the fence, and so develop respect for it. During this educational exercise, "bait" the fence by hanging cornstalk leaves or dainty bits of clover on the wire. One bite will teach almost any cow. Electric fencing is sometimes dangerous to use with horses. They have a more delicate nervous system than cows, and seem to react more violently to a shock. There are even some reports of horses dying of the shock effect. Unless they are educated gradually, with some space to run, they may wheel and charge through the fence after getting a shock.

The electric fence can also be used, though less satisfactorily, for sheep and pigs. For these smaller animals two wires are used, one of them being the ground. Baiting the wire is essential, or they will go right through before learning to fear the shock.

In addition to a stout line fence, and adequate fencing between fields, I feel that every farm should have at least some small stretch of wooded fencerow. It can be along the road or against the woods, but there should be some place on the farm which will shelter wild life and birds, to give joy to your children and yourself. It can be a place where you haul stones from the fields, to be piled into a rough wall. I know the current shibboleth is for "clean fencerows" to cut down insect pests, but I often wonder whether the old rail fences of our ancestors, eating up a rod of tillable soil, to be sure, did not harbor enough birds to keep down the bugs so that spraying was unnecessary. It would be an interesting—and pleasant—experiment to find out.

ADDITIONAL READING MATERIAL: *"A Sand County Almanac,"* by Aldo Leopold.

Free from U.S. Dept. of Agriculture: *Young Apple Orchards, Estab-*

lishing and Management, F 1897 (similar pamphlets for other fruits);
Nut Tree Propagation, F 1501; *Growing and Planting Hardwood Seed-
lings on the Farm,* F 1123; *Woodlands in the Farm Plan,* F 1940; *Meas-
uring and Marketing Farm Timber,* F 1210; *Equipment and Methods for
Harvesting Farm Woodland Products,* F 1907; *Growing Fruit for Home
Use,* F 1001.

From Supt. of Documents: *Establishing and Managing Young Apple
Orchards,* 10¢, Cat. No. A 1.9:1897; *Nut Tree Propagation,* 5¢, Cat. No.
A 1.9:1501; *Apple Varieties and Important Producing Sections of the
U.S.,* 10¢, Cat. No. A 1.9:1883; *Growing Cherries East of Rocky Moun-
tains,* 10¢, Cat. No. A 1.9:776; *American Wood* series (virtually all trees),
5¢ each; *Home Fruit Garden, East Central and Middle Atlantic States,*
5¢, Cat. No. A 1.35:218; *Northeastern and North Central States,* 5¢, Cat.
No. A 1.35:227; *Northern Great Plains,* 5¢, Cat. No. A 1.35:222; *Pacific
Coast and Arizona,* 5¢, Cat. No. A 1.35:224; *Southeastern and Central
Southern States,* 5¢, Cat. No. A 1.35:219.

-19-

WEEK-END FARMER AND
ABSENTEE OWNER

Even a week-end farmer can make his land self-supporting.
Three kinds of livestock, for extra income.
Field crops and a long harvest season.
A muskrat or fishpond.
Land-cleaning, and "let alone" beekeeping.
The absentee owner can operate in three different ways.
Tenants and sharecroppers produce better profits than owners.
Don't "load" the tenant with specialties he does not care for.
Turn the sharecropper's profits into livestock, to keep him working.
A few "don'ts" for owners with sharecroppers.
The farm management organizations can carry the burden for you.
Five factors which tell you whether you are succeeding.

AT FIRST blush there may not seem to be much which a week-end farmer can do—I mean the man who has a few acres in the country, too far for daily commuting, yet within reach for Friday night to Monday morning excursions. Actually, however, he has literally dozens of farming enterprises from which he can choose, nearly all of which, if wisely managed, can make his land self-supporting.

To be sure, some of the major farm enterprises are not for him. Dairying and poultry raising, or the cash crops like strawberries and other small fruits, which require daily attention, cannot be undertaken. But livestock is not entirely out of the picture—provided the week-end farmer has a co-operative neighbor with a view of his land, who will keep an eye on things while the owner is gone. Beef cattle, young stock and sheep are three enterprises which the week-end farmer can engage in without too great difficulty, if he has pasture land, good fencing, and water.

Suppose he wants to go into beef cattle raising. He can purchase young steers, say at 300 to 400 pounds apiece, in the spring, and let them run in his pasture, provided they have water and shade. If there is no spring or brook, an automatic water supply can be arranged, either with a controlled pump or a free-running windmill with overflow drain from a tank. The animals will require no artificial shelter, and in the fall, when the week-end farmer is ready to close up for the winter, he can sell his steers to a farmer for dry-lot feeding. Each steer should have gained about 200 pounds during the summer, and with no winter feed to be provided, the week-end farmer can crowd one and a half to two steers per acre, depending on their size.

Renting pasture for young stock is another possibility. Many farmers, short on pasture of their own, like to farm out their young heifers and replacement stock for the summer, saving the pasture near the barn for their milk cows. If you have fields with shade and water, you will very soon have offers of more young stock than you can handle. The only danger is in permitting the land to be overgrazed, with consequent damage to the seeding. The safest way is to figure not more than one and a half head of young stock per acre, assuming they are all about one year of age.

With either beef cattle or pasturing young stock, the job of the week-end farmer will be to see that the land is not being overgrazed, that the water supply is adequate, and that the fences remain in good shape. Attention to the fencing will be more necessary with young stock than with placid beef cattle. No week-end farmer, however, should try beef cattle or pasturing young stock unless he has made prior arrangements with a friendly neighbor to keep an eye on them during the time he is gone. The promise of a steer to butcher in the fall—either free or at reduced price, depending on the size of the acreage, ought to form a satisfactory barter arrangement. Remember that your neighbor must stand ready to shoo off marauders, act as round-up specialist if the animals break out, and keep enough of a superintending eye on the cattle to note whether any get sick and need attention, hence the job is worth some compensation.

Depending largely on the condition of the fencing, sheep are another possibility. Late-summer and fall fattening of spring lambs, either natives or Westerns, can be carried on if woven wire fencing is in excellent condition, and predatory dogs are scarce. (See Chap. 10, "Sheep Can Be Profitable.") With good pasture you can run as many as a dozen or fourteen sheep per acre. Your neighbor can give you a good estimate of what the pasture will stand in the way of grazing.

Suppose your land is too far from a neighbor, or other conditions are not right for grazing animals. If the land is capable of being tilled, a great variety of field crops may be raised, assuming that you will time your vacation for the harvest period, or plant crops which may be harvested over a long period of time. Corn, flax, small grains and hay are among the possibilities, as well as potatoes, vegetables, squashes and pumpkins. In view of the destructive habits of small boys, melons are probably not a good bet unless you can be there during the ripening season, which also requires rather close attention. (See Chaps. 13, "Where the Garden Can Help Most," and 14, "Cash Crops and Truck Farming.")

If the land is marshy, consider fencing it for a muskrat pond, for geese or frogs, or even for a fishpond, which can be as productive, and certainly as entertaining, as any other acre.

Even if the land is stony or hilly, the possibilities are still good for some income-producing activities. A fruit orchard, nut trees, cutting of pulpwood, fence posts and fireplace wood can be carried on over even the most unlikely ground. (See Chap. 18, "Orchard, Woodlot and Fencerows.")

Suppose, on the other hand, that your acreage abounds with stones, undesirable trees and shrubs, and suffers from erosion. You can spend many a profitable week end clearing the sections suitable for cultivation, plugging up the gullies and improving the rest, until you have made something like a garden out of a wilderness.

Another excellent activity for the week-end farmer, which can be carried on singly, or in connection with any of the other enterprises mentioned here, is bees. Under the "let-alone" system of bee-keeping outlined in Chapter 17 ("The Hum of the Hive") you can have a few or dozens of hives, caring for them on week ends, and leaving them alone during the winter.

The week-end farmer, of course, must reconcile himself to the fact that he is not going to make much money at his farming—but he will not be expending much effort, either. He will take his pleasure in the joy of doing—and he will not be too surprised or crestfallen at the occasional petty thieveries which are the lot of absentee landlords. Good relations with farm neighbors will help to keep these losses to a minimum.

What of the man who does not wish to live in the country, but still wants to buy a farm as an investment, a hedge against inflation, for sentiment, or any other reason that seems good to him? There are certain guideposts which can help the absentee landlord to produce, not only 5 per cent return on his farm investment, but additional cash as well.

Any man shrewd enough to accumulate excess cash for investment in a farm is also likely to be prudent in its purchase and management. He can operate his farm either with a tenant, with a sharecropper, or leave its direction to one of the farm management individuals or companies which have sprung up in recent years. Each method, of course, has certain drawbacks and advantages.

With a tenant, where the land is rented at so much a year, the tenant usually provides all the machinery and livestock, fertilizer, etc., and the owner gets a fixed sum in return, less taxes. Theoretically, the owner need not go near his farm. Actually, of course, by his presence and tactful suggestions, he can do much to keep land and buildings in shape.

A sharecropper calls for more personal attention from the owner, if both are to thrive. A sharecropper usually provides all of the machinery, and sometimes half of the livestock, though this latter is by mutual arrangement. Owner and sharecropper usually each bear half the costs of such items as seed, fertilizer, additional stock, and hired operation expenses like threshing and hay baling. The owner, in addition, pays taxes and building repairs or additions. Both share in the cash sales of farm produce, and in the normal increases in livestock. Until recently owner and sharecropper shared equally in the benefits, but the postwar increases in cost of machinery, feed supplements and the like have thrown an extra burden on the sharecropper, and he is likely in the future to demand a larger share of the returns.

Since the financial success of the farm enterprise depends on the energy, skill and general farm knowledge of the tenant or sharecropper, the exact percentage division of gains is much less important than the capability of the sharecropper. If he is a shiftless incompetent, the profits will be negligible, and there is a good chance of loss. If he is smart and active, your total gains are likely to be greater if he seems to be getting a break on the division, for that will give him a greater incentive to produce.

Take advantage of the normal farm occupancy and ownership cycle. Sharecroppers at their best are young fellows, usually just recently married, who look forward in a few years to owning a farm of their own. They have the enthusiasm and energy of youth, and it is up to you to turn it to the advantage of both of you. In other words, accept the fact that your best sharecropper will not be with you long, and the better he is, the shorter will be his stay. If you help him on the road to farm ownership, he will help you by making the farm a profitable venture. The old-timer, who has never got up enough gumption to own a farm of his own, is not likely to do any better for you than he has for himself.

One of the best ways to help the sharecropper to help himself is to turn as much of his profits into livestock as possible, rather than into easily dissipated ready cash. In this way he will see his investment growing, and will retain his energy and enthusiasm, which might not be the case if his gains went into a new car or other non-productive outlets. He will be ready to leave you that much sooner, but meanwhile he will have benefited you financially more than an indifferent sharecropper would have done, and the example of your treatment of him should give you good candidates from whom to pick a successor.

The temptation, with a good sharecropper, is to crowd or persuade him into unprofitable ventures. You see him raising good crops, or handling a herd well, and it is natural to conclude that he could take on a few more projects. The fact is, however, that few men are good in several branches of farming, and most are good in only one or two. Seldom will you find a man skilled at raising crops who is equally able in a dairy barn. Inclination plays a large part, too. If the sharecropper doesn't like chickens, but you persuade him to take on five hundred or so, and build an expensive chicken house for them, they are likely to show a loss. In other words, study your sharecropper to find out what he does well, and have him specialize in that. Those are the fields where profits lie, and any side excursions against his will are practically sure losers.

A few don'ts for owners with sharecroppers:

1. Don't ever buy machinery, or you will be buying trouble. Let the sharecropper furnish the machinery, or if he can't swing it financially, make him buy part of it, and let his profits go to paying you out. If the sharecropper owns the machinery, and has the duty of keeping it in repair it will be running, and not just running you into debt. A possible alternative is to give the sharecropper use of old machinery which may be on the farm, with the understanding that the machinery stays with the farm, and it is up to him to replace any pieces that no longer run.

2. Have a written contract, with the duties and benefits of each party fully set forth.

3. Don't waste your money and the sharecropper's time on new farm enterprises in which the sharecropper has no real interest. They will probably not pay out.

4. Be reconciled to the fact that good sharecroppers are soon

going to be good farm owners—and possibly neighbors. Help them along the road.

5. Don't "sandbag" the tenant by demanding things like half a hog for your freezer, or a few dressed chickens or bags of potatoes, unless these are figured in the contract. The cash value of such items is really very small, and the ill-will they can generate will cost you much more.

The third method of absentee ownership is through one of the farm management organizations or individuals. The county agricultural agent, agricultural college, farm auctioneers, and the local branches of big implement companies can all direct you to such persons, and give you some indication of their effectiveness. The farm management expert is one more person to be cut in on the total profits of the farm, but this may be more than offset by the advantage of his technical knowledge.

The farm management expert will see to getting a tenant or sharecropper, keep the soil fertile and the buildings in repair, counsel on the most profitable crops and land use, and keep a general managerial eye on the enterprise. So many factors, such as soil fertility and the ability of tenants and sharecroppers, enter the picture that no generalizations can be made on whether you can make more profits with a farm management expert than you could without him. However, the wartime period when practically anybody could make money at farming seems to be over. Maybe you can make more on your own, with close attention to the farm enterprise, and plenty of luck. If you are totally ignorant of farming, however, the farm manager will at least assure you of breaking even, and probably produce a small profit on your investment.

For the novice, just starting out and uncertain whether he is on the right track, a farm manager may be a prudent investment for the first year at least. It would be a means of taking out insurance on breaking even. For the man who wants to jump at once into full-scale farming himself, incidentally, but does not wish to go through the apprenticeship of tenantry or sharecropping, a farm manager as counselor offers the best chance of success.

Curiously enough, the tenant farmer apparently does better than the owner-operator in producing a total profit on the farm. A study of nearly five hundred farms, carried on for sixteen years by the University of Wisconsin College of Agriculture, indicated that tenants showed a better percentage of profit than did farms operated by their owners. (Possibly some of the advantage may be due to the fact that tenants are usually young fellows, able to devote more energy to the farm operation.)

The same study showed five factors which together go to make the farm enterprise a success or a failure, at least as regards dairy farming. The first of these was the number of crop acres, or land under cultivation. Where the acreage was below average, the income was too.

The second factor, crop values per crop acre, also bore a relation to the total net income of the farm, indicating high or low soil fertility, and good or bad crop management.

The third factor, pounds of butterfat produced per cow, indicated whether the herd was good or poor, while the fourth factor, the relation of livestock returns per $100 worth of feed used, showed how efficiently the dairy barn was being operated.

The fifth factor, diversity, covered the extent to which the farmer branched out into the various divisions of farming, such as combining pigs and poultry with dairying.

Curiously enough, the farmer who was just a little better than average in most or all of the five factors, but not at the top in any, showed a greater net return for his labors than did the specialist who stood at the head of the list in one or two factors. Farm management counselors or state agricultural colleges can show you how to set up the accounts which will enable you to determine how you are doing on the five factors.

ADDITIONAL READING MATERIAL: (from Supt. of Documents) *Part Time Farming*, 10¢, Cat. No. A 1.9:1966; *Better Farm Leases*, 15¢, Cat. No. A 1.9:1969; *Farm Lease Contract*, 5¢, Cat. No. A 1.9:1164.

PART III

Where and How You Can Do It

-20-

CHOOSING THE LAND—AND THE COMMUNITY

Check to be sure you are locating in a pleasant neighborhood, and not in a gun-toting feud spot.

A reliable real estate agent can lead you quicker to the farm you want than you can find it yourself. He knows people with money who may help finance it.

Profitable farms cost plenty. Run-down farms call for supplementary income for ten years before they begin to pay.

Good farming calls for good roads.

Look to the fields first for value, because they support the farm enterprise. Watch out for "outlaw," overbuilt and show-place farms. Avoid farms that lie on both sides of a road. Big farmhouses are a liability, unless they can be subdivided.

Fences, paint, gullies and weeds are four vital clues to farm value.

The nearer to town and the smaller the farm, the more you pay per acre.

Insist on a real estate survey, or you may be paying for more than you get.

DON'T forget the intangibles when you set out to buy your farm. You can't get a loan on a view, but you can get a lot of satisfaction out of it, and often it costs no more. And there are certain other nonbankable aspects of the community which can make country living a pleasure or a curse.

For the same reason that you would not knowingly buy a city dwelling, no matter how well built, next to a railroad roundhouse, or in the midst of a bootleg and gangster nest, don't buy a farm just because the buildings are good and the land is rich. The spirit of the community has a lot to do with ease in getting your work done, as well as the joy of living. I know of communities in my own state where the "neighbors," if they can be so called, tear down each others' fences, literally poison cattle wells, and keep the attorneys fat with feuding lawsuits. Anybody buying into such a settlement—and the land is likely to be surprisingly cheap—is just looking for trouble.

Such areas are rare, but it is well to know that they exist. A good spirit in the community actually increases the price of the land, because people do not willingly move away from a friendly place, and hence are loath to sell. That difference in price will be worth paying for. You don't need seventeen college sociology professors with a flock of questionnaires and statistical tables to find out whether the farm you are thinking about is in Happy Valley or Hell's Half Acre.

Any storekeeper in the nearest village can tell you whether the merchants sponsor periodic farm and city get-togethers, and whether they are well attended. The same man can also tell you whether there is any organized recreation program for the youth of the whole area. A third simple check is to drop in at the office of the weekly newspaper and ask to see the files. If the paper carries a lot of items which indicate joint village and country activities, you are in a good neighborhood. The local editor can also answer you on the recreation and farm-village special days, and will talk more freely than banker or merchant about any possible feuding.

Religion is another angle to be considered. If the nearest settlement has just a few houses, dominated by a single large church, flanked by a church school, it means that the prevailing religion of the farmers in that area is of that church. If you happen to be of the same faith, you'll get on fine. But if you aren't, Christianity being occasionally what it is, you may find difficulty in getting neighbors who will help you put in your crops, and your children will go to public schools which are starved and slovenly. For the outsider in such a one-religion community, social life is also sharply restricted.

There is a special joy in discovering for yourself the land you want to buy—but don't pass up the help of the real estate agent, who can be a great time-saver for you. Half a day with him will give you a good idea of what your money is likely to buy in that neighborhood, and whether there is anything in it you like. He is likely to show you two or three very good farms in the price range you are considering—and several very poor ones, to make the good ones look better. Don't be afraid to confide in him, for unless he knows your price limits and the type of farming you are thinking of, he will be working in the dark, and wasting the time of both of you. When you tell him your price range, shave it a couple of thousand, because he will undoubtedly show you farms right on the line and a little above. It's the nature of the beast.

A good farm real estate man is not out to beat you or the owner of the farm he is selling. He wants to make a deal that will stick, and that will be satisfactory to both parties. Unless he does, he will soon be out of business. For that reason, he needs the facts about your economic capacity and the kind of farming you want to do. I will concede that there is an occasional bad egg among the breed, and if you are unfamiliar with the territory, you will need some help in picking out a reliable real estate agent. Ask the local banker and the insurance company operating in that territory for their advice as to who is an agent who will treat you right.

One of the chief advantages of a real estate agent is that he will have on his list a lot of farms which are not publicly advertised for sale. Sometimes he can act as a go-between for a farm that you have your eye on, but which the owner has not previously considered selling. If you were to approach the owner yourself, the price would probably leap. Another advantage of the real estate agent is that he can often arrange for financing. He is in touch with persons interested in farm investment who will take a bigger chance than any bank, insurance company or federal loan agency. Because his commission depends on completing the deal, the real estate agent often sets up the entire financing of the farm purchase.

Whether to buy the same acreage in a good or a poor farm depends on the cash you have and whether you are going to rely on the farm immediately for all your income. The run-down farm takes about ten years, usually, to put in shape. You will require that length of time to bring back the soil fertility, cure gully erosion, rebuild fences, and put the buildings in shape. Meanwhile, there is very little income, and a lot of outgo.

In farming, as in so many other ventures, the old rule still applies: "You get about what you pay for." Some cynics modify this to read: "You never get more than you pay for, and sometimes a lot less." The run-down or neglected farm may be a good buy if you have other sources of income on which to keep going until it produces, if the buildings are such that you can take your time about repairing them over the years, and you want to keep your farm investment low. Don't forget, though, that it won't produce a good income for many years.

Sometimes a farm is low-priced because it is an "outlaw." You are likely to find them at the dead end of a side road, or up a steep hill, away from the main road. In northern climates, they are the last to be "plowed out" in winter—and sometimes they never get plowed out. In very bad weather the farmer has to haul his milk to the main road, because the milk truck refuses to attempt his road. Telephone and electric lines may have ignored him, and can be put in only at great expense to the owner. If he runs into difficulties or emergencies, such a farmer is practically helpless, for no one can or will get near him. An outlaw farm is no bargain at any price, even though land and buildings may be in good shape—unless you intend to be a hermit, for which it is ideal.

In the old days of farming, two wheel tracks were good enough. The farmer stayed home when the weather was bad, and only took the buckboard to the creamery once or twice a week, after separating the milk and feeding the skim to his hogs. But the revolution in agriculture has made good communication roads to farms an essential. The artificial inseminator and the veterinarian are frequent callers at the dairy farm. The parade of corn-picking wagons in the fall, the hayracks for threshing from the field, the utility linesmen, and your friends and neighbors, all need a good road to your farmyard.

There's no use laboring the point. You need to be on a good road, or you can't farm efficiently. For the part-time farmer, dependent on getting to a city job, a good road, always plowed out, is vital. If you are going to keep a city job while engaging more or less extensively in farming, measure distance from your place of work in minutes rather than miles. Drive it during the hours you will ordinarily be traveling to and from work, and check your time. There may be special traffic conditions which will make another location, farther away, but shorter in minutes, more desirable.

One easy way to be sure that you will live on a well-maintained and plowed highway is to buy a farm on the same road where one of the township board members lives. There's a sour old country saying which goes: "You can always tell where the town board members live by looking at the roads." Of course, this may involve electioneering to see that the same crew stays in office. If that fails, you can run for the town board yourself!

If you have children, you'll want a good road for the school bus and a good school for the children. Rural schools vary all the way from one-room rural schools with outdoor plumbing and a sometimes poorly trained teacher up to the consolidated wonders with gymnasium, big library, and nearly everything that a city school can offer. The small rural school is just about as good as the teacher. It can be pretty good, or pretty awful. The big consolidated school districts, taking in a lot of rural territory, are usually wonderful schools, but sometimes they place heavy burdens on a farm, if there is a big bonded debt to be paid off rapidly. There are many instances of school taxes in such districts running higher than the annual rental value of the farm, so make sure what school district the farm is in, and how much the school taxes are.

Of course, you don't need to be on a good road, near schools and markets, with the electric high line running by your door and a telephone in the kitchen. You can head back into the woods on a dirt track, buy cheap land, and pioneer. But that's just what it will be, and you'd better be quite sure you really want to be a pioneer. Incidentally, your production per acre, without the modern helps developed for agriculture, will also be just about that of the pioneer —namely, very low.

Let's assume that you intend to farm efficiently and comfortably, and therefore have picked a desirable neighborhood. You have looked at several possible locations on good roads flanked by the necessary utility poles, and have a general idea of land prices in the neighborhood. How can you pick the farm that will best suit your purpose and your purse?

Look first to the fields, for it is the land which supports the buildings and the farmer. The buildings can help you produce well and efficiently, but even the best buildings will not overcome poor land or a bad layout of fields. The prettiest farms are those nestling in a narrow valley, or folded among hilly slopes. Beautiful woods crowd near the farmyard, with the fields far beyond them. Perhaps a gurgling brook meanders diagonally across the acreage. Only the bottom lands are tilled, because the slopes are too steep. The fields are narrow, crooked, or triangular in shape, and sometimes may lie in three or four separate little valleys.

It's beautiful, all right, but it isn't farming. You will be purchasing—and paying taxes on—something that is not suited to producing good income, or even much income above expenses. If you have to chase your livestock through the woods to the fields beyond and then back again, and haul manure and harvested crops for a half mile or more from the farmyard to the edge of the fields, you won't have time for much productive farm work. You'll just be a traveler. Leave such farms to the man primarily interested in hunting and trapping.

That little brook may flood in spring, making you late in getting your crops in, or ruining the harvest in the fall. Possibly its worst effect is to chop up your fields into areas that are hard to work. Triangles can be plowed, harrowed, seeded, cultivated, mowed and harvested with machinery, but you can waste a lot of time on the turns, doing it. This type of field also means considerable extra amounts of fencing.

Besides the likelihood of erosion, a hilly farm also presents practical dangers in farming. Literally dozens of farmers are pinned and killed each year under tractors that tip over on hillsides. Even if you escape fatalities, there are always added difficulties on a hill farm, such as extra power needed in hauling manure up the slopes, and getting a load of hay safely down. Such land is often better suited to trees, than to farming.

Watch out, also, for farms split by a highway. A road cutting diagonally through a farm, as many of the new superhighways do, can carve your fields into triangles just as effectively as a stream, and take up just about as much land. A road in the middle of your property also means a perpetual traffic danger. In the old days of slow-moving horses and wagons along a dirt road, a farmer could have his barns on one side of the road and his pasture on the other without much risk, but if you try to weave a herd of cows through a fast-moving line of cars, you may lose a cow, as well as your temper. True, you have the right-of-way with stock, but the cow will be just as dead after the car hits her.

The same difficulties and hazards occur when you are threshing from the field, hauling in hay, or doing any of the other farm tasks which involve transport of animals or machinery. In other words, your farm will be a lot more usable, as well as safer, if all of the land and buildings are on one side of the road.

Long rectangular fields, rather than square ones, are the most efficient for modern machinery. The "headland," or turn-around space needed at the end of each field to swing machinery back for the next row, will raise some crop, but it naturally won't be as productive as the rest of the field for crops which are cultivated during

growth. Since it takes about eighteen feet of headland at each end
for a turn-around, you will cut down this less productive space, and
save more time because of fewer turns, if your fields are longer than
they are wide.

Note whether all the fields can be reached by one short, central
lane. The longer the lane, the longer the double fencing, and the
more cropland taken out of production.

Walk around the fields, wiggling an occasional fence post to see
whether it is solid and not rotted near the ground. See if the fence
wires are taut, and staples driven in snug. A trip through the fields,
particularly in summer, will give you a good idea of the soil fertility.
Thin stands anywhere mean poor soil, and on slopes spell the effects
of erosion. Rolling land can be fertile, but if the slopes show signs
of erosion, it means you will have to do strip cropping and contour
plowing, which will restore the fertility and stop erosion, but take
you more time—and maybe more fencing—in the doing.

The owner should be able to tell you the crop history of each
field for at least five years. If he has been alternating corn or grains
with clover or alfalfa, which restore nitrogen to the soil, he has been
practicing a good rotation to keep the land in shape. Corn on the
same field more than once in four years, for instance, probably
means that the land is being robbed, unless heavy amounts of extra
fertilizer have been applied.

When you look at the buildings, consider whether they are suitable for the kind of farming you intend to do—and whether the
farm is "overbuilt." Large, expensively constructed barns and other
buildings are fully reflected on the tax roll, and if they are bigger
than you need, they will be cutting down your income, through
extra taxes, extra investment, and extra maintenance. Such overbuilt
farms are frequently on the market, simply because they are difficult
to operate at a profit. These observations apply equally, of course, to
"show-place" farms, built by some rich man as means of beating the
income tax collector, or to please his vanity.

Dig your pocket knife into the beams in the basement of the
house, and the lower section of the barn, to make sure they are solid,

and see whether the foundation walls are cracked and heaved. Look at the paint job on house and barn, and note especially whether the window frames in the house are in good condition. Too often they are so rotten that the glass is ready to fall out.

Through a natural enough progression, farmhouses are often too big for present-day use. The pioneer may have started with a sod hut or log cabin, but his grandson, with the farm cleared and paid for and money rolling in from good crops, frequently took a look at his many children and built a house big enough for them all, plus the hired man. Sometimes he got the house done just in time to see most of the children decamp for an easier life in the city.

Unless you have the family to fill it—and keep it filled for several years to come—a big farmhouse means extra heating costs and maintenance. However, some of the large old farmhouses can easily be divided to make room for another family, and thus can be a source of income.

If you go around with a real estate agent, he is likely to point out that the land is worth so much, and the buildings so much—and he'll probably be right. But the point is that if you want to make both ends meet in farming, you should "buy the land, or buy the buildings, but not both." Farms are usually sold at so much per acre, with the buildings thrown in, and this is a realistic recognition of the fact that the land supports the buildings.

When you have obtained a general idea of the prevailing prices of farmland in the area you have chosen, here are four simple tests which you can make yourself in less than an hour, to determine roughly whether the price of the farm you want represents good value. These tests are: fences, paint, gullies, and weeds, all of which indicate whether the farm has been kept in good shape as a going concern.

It is hardly necessary to point out that good fences mean a careful, thorough farmer, usually with enough livestock to insure plenty of manure on the fields. Well-painted buildings, especially if the paint is old enough so that you are sure it was not just recently applied to help the sale, are a similar indication of good husbandry.

Bad gullies, on the other hand, show sloppy farming. They are an indication that the richness of the land has been mined rather than maintained. While gullying can be cured, it means that much of the priceless topsoil is already gone, and much expense lies ahead to stop the gullies and restore fertility.

Weeds, like gullies, also indicate a run-down farm. Large patches of thistles in a field, for instance, show poor farming practice, which your time and money will be spent to correct.

It is advisable to have the fertility of the soil in each field tested (which can be done through the county agricultural agent), but there is also a simple test you can make yourself, without even going into the fields. If you see large patches of "sour dock," a low weed with a light yellow flower, it is a good sign of a poor field. This plant is usually to be found only on thin, worn-out soil, and the large yellow masses can easily be spotted hundreds of yards away. On the other hand the thick-rooted yellow dock, a much taller weed, avoids thin soils, and is usually found in bottom lands and other rich soils. It is a good sign of fertility, even if large quantities of them also indicate a lazy farmer.

There are two other general indicators of price: one is size of the farm, and the other is nearness to a good-sized town or city. The smaller the acreage, the more you will pay per acre. This is so because things like the cost of a well, driveway, buildings, and other improvements outside of the fields, must be spread over a smaller number of acres. Nearness to a city also affects the price of small acreages, because of the ease with which they can be sold for subsistence farming, and the speculative element of possible future subdivision.

An important point: at the time you make your purchase agreement, specify in writing that you are buying a definite number of acres, for the specified price per acre, and that this purchase will be *as shown by survey*. In many parts of the country either the original government surveys were carelessly made or the fence lines were put up haphazardly by the original settlers. A man may honestly believe that he is selling you a 120-acre farm, but a survey may show that it actually contains only 105 acres. Your line fence may be exactly on the line, but it could also be as much as 50 feet or more inside your neighbor's land or outside the land that is rightfully yours. Where such discrepancies are shown, the fences can be moved, or they can be left as they are and adjustments made in the purchase price. Real estate contracts usually provide that the land sold is "120 acres, more or less," which gives you no recourse in case an error is later discovered. But you don't need to sign a contract like that. You can insist on getting exactly what you pay for. The few dollars that a survey costs may also be well worth it later, if you should decide to sell the farm.

-21-

WHERE TO GET THE MONEY

The size and value of the farm you can buy depends on how much cash you can raise.

Mortgage companies lend on long-term "true value" of cropland and not on pretty views.

Federal land bank offers best bargain in mortgages. Don't forget relatives and friends, if other sources fail.

A mortgage on a small farm depends on your city pay check rather than crop production.

Land and buildings are only part of the cost. Machinery and stock will run to an equal sum for a dairy farm, somewhat less for a grain farm, but nearly 100 per cent loans can be obtained for this personal property.

"Feeding out" finance companies help you swing most production projects.

HOW much cash will you need to swing the land you want? The man who can lay down the full purchase price for a good-sized farm—and have an equal amount left over for stock, machinery and other expenses until he gets into production—can stop reading right here. The good Lord has already taken care of him better than any advice he will find in this book.

The farm you pick and the cash in your pocket go hand in hand. Since every case will be different, the best that can be done is to set up certain general signposts which will indicate how much you can bite off in the way of farmland without choking on a wad of bills.

The chief thing to remember is that mortgage and lending companies have been burned so often on bad loans that they have evolved a set of principles to make sure it doesn't happen too often again. They like you personally, and they want your interest money, but their primary concern, naturally, is to make sure they get their investment back. They know, from long experience, that a farm is worth what it will produce in the way of animal and field crops— and it isn't worth anything more. This is a lesson that some beginning farmers—and some old hands at it—have to lose a farm to find out. This rule on cropland values, however, does not fully apply to small farms—say under fifty acres—which usually find a ready market as dwellings or for subsistence farming, and hence are not considered entirely as farm business enterprises.

For the larger farms, the lending representative may be enthusiastic about the view or the way the shadows make exciting patterns in the woods, but it is a pure excursion into esthetics for him. What he really cares about is land under plow, for that is the only thing which will produce income to pay interest and principal. When it comes time to make out the papers, he isn't lending on view, or marshland, or on pretty woods. From his standpoint, these elements which may make the farm especially attractive to you are just so much extra burden of taxes and fencing. Even a good stand of woods does not impress him, for he is aware that if the woods were really as good as you think they are, somebody would have logged out the good timber long before this.

The mortgage company isn't even going to consider present market value, except as one element in a variety of considerations. The agent takes what he calls the "true" or long-term value of the plowed land as the basis for his loan. If the land is the kind that will produce eighty bushels of oats per acre, he considers the price of oats for the last twenty years, remembering the extreme lows as well as the occasional highs, and figures what the average cash return would be over a long period. He feels safe there because he knows that you—or your successor—will be able to pay out a loan with that basis.

But the banker isn't going to lend you the full "true value." Just how much of the true value he will lend depends on the agency. Probably your best bargain in financial agencies is the Federal Land Bank, which requires only 30 per cent of the true value paid in, and will give you a loan to pay the other 70 per cent. Even more important, the Land Bank will give you a mortgage running for thirty years, with "level" payments. In other words, you pay the same amount, year after year, until the loan is retired, rather than making heavy initial payments, when your crops and livestock are not yet producing at capacity.

Insurance companies are ordinarily limited by law to an investment of 50 per cent of the true value. Their loans are usually not for as long a period as those handled by the Federal Land Bank. On the other hand, the local bank in the area where you are thinking of buying, will often lend up to 60 per cent of the value. The disadvantage is that the local bank doesn't want its money tied up in a long-term mortgage, and may lend for a period as short as five years. The difficulty here is that you may find yourself at the end of five years in a period of low farm prices, when no one wants to put out any money on farms.

There remains one other financial possibility: relatives and friends. If you can't make the big down payments required by the financial institutions, you may be able to swing a loan from Aunt Maria, if she likes your character and industry. Even total strangers, not dependent on a steely-eyed bank examiner for continued existence, have been known to plunge on a farm mortgage, figuring that they can get their money back eventually, even if they loan as high as 80 or 90 per cent of the true value. Here is where a real estate agent can be of real help. He usually knows of several individuals interested in farms as a speculation.

But suppose you are not buying a large farm, and merely want a small one for part-time or subsistence farming. Your chances of getting a loan in such a case are dependent not so much on the land, as on your city pay check. After all, if you can satisfy the lender that your income from other sources will be enough to pay your expenses and still pay off on a loan, you have a good chance of getting the money. He knows that even if your strawberry cash crop fails, and the insects get all the rest of your crops, you will still be able to keep up your payments.

The smaller the farm, in fact, the more the loan will depend on your outside income. This is so because the cost per acre rises as the farm gets smaller. And the smaller the tilled land, the less net income there is to be produced from farming. Under these circumstances the lender will be highly curious about the steadiness and chances for future advancement in your job, and indifferent to the kind of soil and the buildings on your farm. There is a hint here that the small, high-priced farm is more a place to live than an income producer.

If you are buying a small farm or just a few acres for country living, you will want to set up the repayment plan on a monthly basis, in accordance with your income. This will avoid the necessity of being stared in the face once or twice a year with a whopping big mortgage payment—and no money to pay it with.

However, if you are buying a good-sized farm, which you intend to be self-supporting or your sole means of making a living, it would be best to set the payments to fall due a month or two after your big crops are harvested and paid for. Thus a man specializing in raising pigs would want his big mortgage payment to fall due soon after the first of the year, because he will market most of his pigs in the late fall or early winter. A grain farmer would likewise set his payments for fall, but a beef cattle fattener would prefer spring.

The time to arrange for these payment schedules is *before* the loan is completed. This is the time when the lender is willing and co-operative. After the loan is once set up, it may be another story to try and get the dates of payment changed. Whether you are going to make the repayments from the monthly income of a city job, or from farming operations, be sure to leave enough margin to take care of crop failures and emergencies. Small payments spread over a number of years are easier to handle than heavy initial payments when you are just starting out.

Don't forget, in considering financing, that the investment in land and buildings is only part of your expense. If you are going into livestock farming, particularly dairying, your investment in cows, pigs, haying machinery, tractors, corn planters, and the like will be about equal to the cost of the farm per acre. If you are going into grain farming, however, without livestock, the investment in machinery will be considerably less than the cost of the farm.

The man planning on dairy farming can expect to make the heaviest outlay. If he has laid down—including a mortgage—$15,000 for land and buildings, he can expect to put up just about the same amount for cows, milk cans, milking machine, cooling tanks and tank-washing outfit, silo filler, haying and corn-cultivating and harvesting equipment, wagons, tractors, combines, and the like, plus a substantial sum for each cow.

Altogether, at current prices, a man planning on a herd of twenty cows, and the land and buildings to maintain them, can count on an investment of between $20,000 and $30,000. For the novice, such an investment would have to be protected at first by hiring a trained operator, which would cut down on the available net income.

However, the picture on financing of livestock and machinery—usually called the chattels—is brighter than it is for financing the land and buildings. Some economies can be made by buying machinery secondhand, either at auctions or from dealers, and it is often possible to arrange for nearly 100 per cent financing of them, so that the cash outlay is very small. This applies also, of course, to new machinery. Nevertheless, it must be remembered that financing charges are necessarily high, and until the machinery is paid for, there will be just that much less disposable income.

Like machinery, the cows, beef cattle, pigs, sheep, even poultry and some crops—anything that grows and can be sold later for more than it costs—can be covered by a chattel mortgage at close to full cost. This supplementary financing, some of it falling under the head of "feeding out" programs, is handled by a variety of agencies.

Local banks are one source of such credit. Others are short-term credit agencies, Production Credit associations owned and operated under federal supervision by farmers, and the financial terms arranged by feed companies.

Suppose, for instance, that you have plenty of hay, pasture and corn, and want to raise beef cattle. In consideration of your equity in the feed on hand, the bank or the Production Credit association will give you the cash to purchase steers. If you buy very young stock, say at three hundred to five hundred pounds, it will be nearly two years before your livestock crop is ready to market, but you make no payments until after the crop is sold.

The same can be done with young weanling pigs, though here of course the term is much shorter, since the pigs will be fattened and off to market in six months, if handled properly.

Maybe your acreage is not big enough for large-scale operations in beef cattle or pigs, but you'd like to take a flier in turkeys or broilers. The large poultry feed companies have worked out balanced rations for the birds, and will advance the feed, taking their payment when the birds are sold. Here again, though, the lender is interested in your equity. He expects you to advance the cost of the poults for turkeys, or young chicks for broilers.

The feed companies are also apt to look somewhat warily on the novice. With a feed investment for turkeys which can run to more than $4.00 per bird before they are marketed, they want to be sure that you know something about what you are doing. The wisest plan would be to start the first year in a small way, and learn the extent of the hazards and labor before you plunge. If you can show that you started with twenty-five turkey poults, and raised them to maturity, without too high a percentage of loss from blackhead, foxes, wet weather, and the other misfortunes that turkeys are heir to, you should have no difficulty in getting a feed company the next year to swing the graining of several hundred birds. This applies also to broilers, ducks, and the like. Try a small sample at first, to learn the ropes and show your ability. The feed company is going to insist on some showing of skill anyway, and you will be protecting yourself from the chance of heavy loss by experimenting in a small way first.

ADDITIONAL READING MATERIAL: (from Supt. of Documents) *Credit Road to Farm Ownership* (with list of other reference material on subject of farm mortgage credit), 10¢, Cat. No. A 72.4:18; *Farm Mortgage Credit Facilities in the United States*, 55¢, Cat. No. A 1.38:478.

-22-

GOOD ADVICE AND WHERE TO
GET IT

Before you hunt up the professional experts, try your neighbor.

Farmers are human—they love to give advice.

*Watch your neighbor's farming methods—you can be a pretty good
farmer, just by copying him.*

*The feed store merchant has bags of wisdom. So does the hardware
man.*

*The county agricultural agent is there to help, and he's a quick man
with a pamphlet.*

*Further resources: the professors in the land grant and agricultural
colleges.*

*Don't waste time reading too many bulletins or books—not even
this one!*

The federal soil and crop agencies, and breeder associations.

LITERALLY thousands of books and pamphlets line the library
shelves to give you the facts on every conceivable aspect of farming,
and the federal government and most states maintain phalanxes of
experts ready to move in with sage advice if you but crook your
finger. Nevertheless, you can and should get most of your information and help from your nearest neighbors.

Books and pamphlets are necessarily written in somewhat general terms, to cover wide varieties of soil and climate conditions, to say nothing of the differing abilities of farm operators. The expert, also, is sometimes inclined to be wordy or cagey or both, apt to set forth the ideal solution at length, regardless of whether you have the means or time to carry it out. More about them later.

Before running to the books and the experts, try your neighbor. Naturally you are not going to put on a front with him, and pretend that you know all about farming. It is a pose that any farmer can see through in a minute, and he would have nothing but derision for you if you tried it. You will go to him strictly as a novice seeking help from an old-timer in the business, and if he is a good farmer, he will practically break his back to help you.

He knows the crop history of your fields, what was planted on each segment for the last several years, and how well it did. He is aware of the local conditions of soil and weather which makes a particular strain of alfalfa a good drought-resister, and one variety of oats a heavy producer in that locality. Check on his advice if you like, but rely on it heavily, for it will probably be the best you will get.

The countryman, during his long years of farming has accumulated an incredible store of practical tips. He knows the best time of year for that area to plant your small grains or corn, when the color of the clover blossoms changes to indicate that it is time to advance with the mower, how many head of livestock a certain field can support without becoming overgrazed.

For animals, your neighbor is also an ever-present source of help in time of trouble. He can tell you what to do if a lamb refuses to nurse, how long to leave the cows in wet alfalfa without danger of bloating, and ways to be sure of getting the greatest number of live piglets when your brood sows are farrowing.

One could go on at length, but the emergencies and occasions for advice will spring up of themselves, and each will be different. You need not feel apologetic in asking for advice. Everybody loves to give it, and farmers are no exception. In the country there is also the added bond of fellowship in a common endeavor.

You will be careful, of course, not to make these pilgrimages for advice just a one-way street. The system of barter is deeply ingrained in the farmer's blood, and while he would resent offers of money for his advice or small assistances, he will expect you to make it up to him in other ways. Things like a box of candy occasionally for the family, the gift of a pair of pliers where you "just happened to buy an extra one," and the like, are the oil that smooths such country business relations.

In addition to asking your neighbors outright for advice, watch their methods too. "A man can be a pretty good farmer just by looking over the fence, seeing what his neighbor is doing, and copying him," the saying goes. The old hand is likely to be good and sound, or he wouldn't still be in the business.

Note especially what crops your neighbors put in, what breeds of livestock they carry. They may not be the superb new jumbo stuff you have just read about in the slick-paper farm magazine, but do not be hostile to them for that reason. It isn't always true, of course, but the chances are that your neighbors, in the slow, careful way of the country, have discovered the types of animals and crops which do well on that soil and in that area. Before you pioneer with a strange breed of cattle or a new strain of pigs, make sure that there is some good reason why your neighbors are all wrong. Maybe they aren't.

"Keeping up with the Joneses" is ruinous in the city, but imitating your neighbors in the country is likely to be just good business sense. If everyone around you has Guernsey cattle and Berkshire hogs, for instance, you fit into the pattern, if you have them too. If and when you run into difficulties in handling your livestock, your neighbor is familiar with the quirks of that breed, and will know what to do about it. A man who has handled Holsteins all his life will feel free to give you advice on your own Holsteins, but might hesitate to stick his neck out about the idiosyncrasies of Brown Swiss or Jerseys. The same line of reasoning applies, of course, to other animals, poultry, and crops.

Raising the same general type of things as your neighbors has still other advantages. You may need the services of a bull for your cows, or a boar for your young gilts. It is easier to get one from your neighbor than to scour the countryside in search of a herd sire of an unfamiliar strain.

Another consideration is the value in indirect ways of conformity. You are probably moving into a strange neighborhood, and although farmers are by nature individualists, they have a streak of suspicion about outlanders who do things differently. It may seem far-fetched to suggest that your children may get poorer marks in school because you have decided to pioneer with Yorkshire hogs instead of the favorite red Durocs beloved of your neighbors, but it could happen. The social pressures in the country can be fully as exacting as those of the city, and your every departure from the normal means that you will be singled out as "queer." If you don't mind, more power to you. This country can stand a lot more individuality. All I am trying to suggest is that the beginner, the new countryman, can make things a lot easier for himself if he starts out doing what his neighbors are doing. After he has established himself, he can branch out into almost any field.

After your neighbors, probably your biggest help in learning the lore of the countryside can be the small-town merchants who serve that farming community. The feed store operator can help you work out a balanced ration for your livestock, suggest the proper fertilizers for the crops you intend to put in, give you tips on care of poultry. Here again you benefit from the universal human love of giving advice—yours for the asking, if you ask as a novice. Make it a habit to "waste" a few minutes chatting with the feed store man, particularly if a couple of other farmers are standing around. I have never failed myself to acquire a few extra nuggets of wisdom just by listening at these chance encounters. The local hardware merchant knows of easy ways of doing farm repairs, if you just tell him your trouble, and the filling station man can give you tips on getting the most out of your farm power equipment.

So much for the amateurs—your neighbors and business acquaintances—who in general have no axes to grind, and who are near enough to you to beware of handing out advice which may boomerang. The professionals fall into a different category, both in the type of information they are likely to furnish, and their availability.

Take your county agricultural agent, for instance, whom you can usually find in an office at or near the courthouse. He is paid partly from federal and partly from local funds, and he is there to help those who ask for aid. Usually he is the product of a land-grant-college four-year agricultural course, has read innumerable pamphlets and books and seen demonstrations of the best ways of farming. He has a stock of pamphlets on hand, and is a walking catalogue of other county, state and federal agencies set up to help the farmer. He can test your soil samples, or tell you where to have it done. He has a map of the soil characteristics of the county, and knows the crops that will do best on each type. He is a mine of information on what sort of rations to feed livestock and other animals, and he is in touch with local quality growers, both of animals and crops, if you are looking for the best seeds or animals.

Many farmers have nothing but disdain for the county agricultural agent as an "impractical idealist," but you will find this feeling is not shared by the more progressive farmers. It is true, of course, that the county agent is interested in promoting ideal methods, and therefore is not likely to be a good source for the short cuts and "lazy farmer" ways of doing things which can produce good crops, but not necessarily the best ones. However, the task of the county agent is to foment the revolution in agriculture by pointing to the best ways of doing things, so he should not be blamed for sticking to his guns.

Often the county agent is stopped on the courthouse steps by an impatient farmer who may ask what he should do, for instance, to control an invasion of grasshoppers.

"Step up to the office," says the county agent. "I've got a good pamphlet by the federal government on how to control grasshopper infestations."

"No, no," the farmer is likely to reply. "I haven't got time to riddle out a pamphlet. Here's the back of an envelope. Just put down what I have to do to get rid of them. That's all I want."

The farmer may be using the right approach, at that. He doesn't care about the history of grasshoppers, and a dozen different ways of controlling them. The county agent can tell him quick about mixing poison with bran and scattering it in the dead furrow near the fencerow, so many hundred pounds per acre. That's all the farmer wanted to know in the first place, and it is to the credit of most county agents that they can furnish an answer right off the bat, right on the back of the envelope.

In many high schools serving rural areas you will find an agricultural teacher—the "Smith-Hughes man," named after the law which set up federal aid for such teaching of vocational agriculture. Like the county agent, he is a storehouse of information, and will be glad to help you on specific problems. Both the county agricultural agent and the high school agricultural teacher spend much of their time in promoting interests of rural youth—giving them instruction in the best methods of agriculture, and in stimulating the recreational and social activities intended to make farm life more attractive to young people. You may find yourself, on short acquaintance, being invited to help out as a Four-H Club adult leader, and it is a job that will endear you to your neighbors.

Back of the county agricultural agent and the vocational agriculture teacher are the land grant colleges, such as the great Midwestern universities, and the somewhat similar agricultural colleges of the Eastern states. If the county agent or vocational agriculture teacher is not easily available, you can call on the professors of the institution in your state for further assistance. The horticulture department will diagnose the ills of your raspberry bushes if you send them a few sample canes, you can get a soil analysis, help on problems of breeding and feeding, lists of seed varieties tested for that state for production and disease-resistance, and countless other services, yours for the asking.

The county agent, the agricultural teacher, and the state colleges of agriculture are literally swimming in bulletins available to you free, on every variety of crop and livestock problem. In addition, the county agent and the college have lists of federal bulletins, thousands of them, on all the major crops and livestock problems, on vegetable and small animal specialties—in fact, on everything.

The chief danger for the beginning farmer is that he also will start to swim in bulletins, which can become a disease. If you are really ignorant of a subject, and can't find quick advice from a neighbor, a bulletin can be helpful—but nothing beats getting out on the end of a hoe or a pitchfork to stimulate production. In other words, don't let the acquisition and perusal of pamphlets divert you from the main goal of producing bumper crops. Don't even spend much time reading this book—just get out there and pitch!

The roster of agencies which can give you information, and occasionally help, is almost limitless, but here are some of the major ones:

The Federal Soil Conservation Service will study your land, help you with the layout of fields, contour plowing, erosion control ditches, etc., to make sure that you keep your precious topsoil. Some parts of the country are organized into soil conservation districts, which have machinery to rent for erosion control work. For the past few years the Federal Government has been making direct payments to farmers who practice good soil conservation measures.

The Triple A (Agricultural Adjustment Administration) is another federal agency interested in saving soil and in cropping methods. It is administered by the farmers themselves, through a county committee, and township representatives. The AAA is interested in how much fertilizer you use, what crops you are planning, and what you are doing about erosion and weed control, for some of which work there are subsidies. The county AAA office usually has an aerial photograph of your farm, showing the cultivated lands.

State agricultural departments are interested primarily in broad problems affecting masses of farmers, but usually they have divisions ready to step in on questions like control of bovine tuberculosis or

Bang's disease (brucellosis or contagious abortion) in cattle, and hog and poultry diseases which carry the danger of epidemics.

The Federal-State Crop Reporting Service, with offices in each state capitol, can give you information on yields, average production, most commonly grown crops, and the like.

In strongly agricultural states there are statewide dairy, beef cattle and swine breeding associations, and often county organizations of the same type. In addition there are national associations for many of the leading breeds. The local and state groups are interested in promoting purebred stock, and can chiefly be of service to the beginning farmer in listing sources of breeding stock.

Similarly, there are state organizations of grain growers, and groups of specialists who raise certified seed.

At the other end of the production picture are the crop and livestock marketing associations, usually co-operatives, which will give you group bargaining power in disposing of such things as wool, livestock, and small truck farm cash crops. Your county agent can tell you which of these groups is active in your locality.

Where to obtain pamphlets:

Pamphlets on hand, usually nontechnical bulletins, are distributed free by county agricultural agents, state agricultural colleges and extension services, and by feed and implement dealers and similar farm suppliers. Under present regulations, each person can receive ten free pamphlets at one time from the U.S. Department of Agriculture, Washington, D.C. 20250. Note, however, that pamphlets which carry a price notice are to be obtained from the Superintendent of Documents, Government Printing Office, Washington, D.C. 20402, even though they are on Department of Agriculture lists.

Because of the literally hundreds of thousands of publications available throughout the government, the Superintendent of Documents has no master list. The agricultural list, "No. 11—List of Available Publications," is for sale by the Superintendent, or can be examined at county agricultural agents' offices and some libraries. It includes instructions for ordering, and is published every three years.

23

FIRST THINGS FIRST

Don't try to put city surroundings into the country.

Save your concrete, at first, for the things that will bring in income.

A long-range plan for your farm's development.

Have a wide entrance drive and fair-sized farmyard—you never know when a boxcar may roll in.

Your wife needs water in the house as much as the cows need it in the barn. Plan a reservoir for fire protection.

An easily attached filter will clear up that cistern water.

A fishpond? Why not? It's supposed to be as productive as any other acre on the farm. And while you're at it, you can make a swimming pool for the kids for a few dollars.

Put in your own sewer system.

Consider giving the old house to the tenant, and building a new one for yourself.

Never tear down an old building until you have thought three times on whether it can be converted to another use.

If all the fences must be rebuilt, this may be the time to rearrange the fields.

Turn your steep slopes into woods—they are no good for pasture anyway.

THEY say in the country: "You can always tell when a city guy buys a farm—he starts running wild with a concrete mixer."

Maybe it's a hangover from pounding city pavements, to want to surround the house with plenty of cement sidewalks, and throw up battlements of concrete retaining walls wherever a grassy bank begins to slope. Eventually you may want such luxuries, but they represent misdirected effort when you are just starting out. What's the real use of a cement walk to the front door, for instance, when everybody and his uncle will follow the old country custom of piling in through the back door and into the kitchen? Of course, if you have plenty of money, and are more interested in appearances than actual farming, you can let the concrete mixer run as wild as it wants, without hurting anything but your own pocketbook.

I got a graphic lesson on the folly of too much concrete mixing when I moved to my own farm. The former owner had decided, without too much reflection, to get his herd of dairy cows onto the city milk market, which meant a milkhouse, cooling tanks, and all that goes with them, instead of shipping to the condensery, where you merely cool the milk by letting the cans stand overnight in the water tank. He had bought a small concrete mixer, made forms for a milkhouse foundation, bought cement, sand and gravel, and poured the footings. And then he decided that the extra sanitary requirements were not worth the effort. So there the foundation sat, a weedy, cumbersome nuisance; not worth completing, and too heavy to move.

Later the idea struck him to go in for bees, but he wasn't going to be content with two or three hives. He started at once to build a honey house, for extraction from combs. In went foundations, and he even made concrete window frames. But before he got past the foundations, the few swarms he had started with contracted foul brood, and he abandoned bees. The honey house foundation, in a corner of the orchard, was another nuisance.

This unfriendly introduction does not mean that concrete has no place on the farm. What I wish to emphasize is that you consider long before you begin, think first of the places where concrete will help to bring in more income. I know that the farmers of my own Midwest have long been justly criticized for putting every spare nickel into improving the barn, with almost nothing for the house to ease the burden of the farm wife, but your first purpose should be to establish a profitable venture.

Incidentally, when you do start pouring cement, consider ready-mixed concrete, trucked to your door and spouted right into the forms. If you have your own sand and gravel, and plenty of time to get it out, you can shave the cost, but if you are buying all the ingredients, ready-mix is cheaper, to say nothing of the labor you are saving yourself. One of its greatest advantages is a uniform mix, likely to stand up through the years, which cannot always be said of homemade mixtures. However, if you have a lot of small repair jobs—putting in a step or two, patching a piece of floor, and the like—a small concrete mixer, powered electrically or by a gasoline motor, enables you to do the work in your spare time.

The dangers of running hog-wild with a concrete mixer point up the advantage of having a long-range plan for the farm development. Every farm and every farmer are different, so no single plan can be laid down which will suit all or even most conditions. However, there are certain general suggestions which will help you to draft your own plan, so that you waste no motions and no money. They may save you from tearing down a building which can be converted later to produce income, or stop you from constructing something that will only be a monument to inexperience.

Let us begin right at the front driveway, and proceed through the farm, pointing out considerations which can help you in planning for the future.

You want, first of all, a good wide entranceway, with an easy grade, capable of handling any-size machine, because farming today flourishes on machinery. You want to be prepared for anything from a bakery wagon to a tractor-trailer stock truck, or you may be taking down a fence and building temporary roadbeds to accommodate some emergency. For the same reason, if you go in for the ostentation of ornamental entrance-gate posts, you will want them at least twenty feet apart—at which distance they begin to look slightly ridiculous, but not so funny as a farmer bringing home a new hoghouse on a flatbed trailer, and blocked outside his own farmyard. The driveway up to the farmyard, of course, can have trees along its sides, but preferably with the same twenty foot clearance.

It may be less esthetic, but I like to see access to the farmhouse from the farmyard, rather than directly from the entrance drive. The reason is that unless you have a separate branch of road up to the door, your driveway will be repeatedly blocked by the cars of friends or the trucks of merchants, just when you are in a hurry to get by with a big load. If all the vehicles don't come to rest until they hit the farmyard, you can get at them easily in case of break-downs, and your main artery for commerce and sociability is always open—even for fire trucks.

The farmyard itself will be governed in size somewhat by the near-by buildings, but it ought to be big enough to hold several pieces of machinery and still leave room for trucks and cars to maneuver. All your main farm buildings should have at least one door opening on the farmyard, to ease the job of loading and unloading stock, machinery, feed, repair materials, and so forth.

At the edge of the farmyard, also, for the easy access by repair-men, should be the central utility pole, with wires to house and to other farm buildings. If your current comes in on the big wires to this pole, you can use smaller wires to all the buildings and have less line loss than if you carried wires through the house and on to the barn, for instance. The importance of a central distribution pole is emphasized, for instance, when your wife is using the electric stove and the washer in the house, at the same time that you are using a feed mill and electric water heater in the barn. If all came from one wire, something would have to give.

It may not seem necessary to discuss the need of an adequate water supply, but there are several angles to consider, depending on the type of farm. Our ancestors got along, with great effort, using a bucket dropped from the end of a rope into a wide well shaft, which was either laboriously chipped through solid rock, or made with masonry walls through dirt. Sometimes they just used a running stream, but usually they located on low land, so that water would not be too hard to get. If the farm has a dug well, the water of course should be tested at the state health laboratory for purity, and the well should be capped with concrete, both for purity and safety. It's a good idea to test the water from any well.

Modern wells are driven, rather than dug, and many states require the well driller to certify to the purity of the water, and the fact that he has capped the well, before he can present his bill. A driven well usually consists of a casing four to eight inches in diameter, down below water level, or to solid rock through which the drill continues until a sufficient head of water is reached. A pipe two inches in diameter hangs in the center of the casing, and a jointed "sucker rod" runs up and down inside this pipe, to lift the water. The casing, central pipe, and a pump on top will get you water, but what force you use to get it up depends on your purse.

The cheapest way is a strong back on a pump handle. Next comes a windmill, which can be disconnected and the pump handle used when the farm wife wants a bucket of cold water during a spell of calm weather. If you have no windmill, or it is broken beyond repair, you can install a pump jack with gasoline or electric motor for less than two hundred dollars.

Because of the enormous gallonage consumed by livestock, you will always find water piped to the barn, even if it isn't to the house. The system may be a simple overhead pipe on stilts from the well to a tank in or near the barn, or it may be piped underground to a hydrant faucet in the barn. (A hydrant faucet is useful in climates where the temperature drops below freezing. The valve turns off below ground, and the water between faucet and valve drains off into a gravel bed.)

If you are playing fair with your wife, so that her end of the work has as many conveniences as yours, you will get running water into the house as soon as you can. Actually it is an economy, for she will do more work if she doesn't have to run out to the well several times a day! One method of getting water into the house is by a gravity system. Put a fifty-gallon tank in the attic or the corner of a second-floor room, and you'll have running water for the kitchen and a downstairs bathroom. To avoid long runs of pipe, the tank should be above the sink and bathroom. A larger tank will hold more water, but maybe the rafters won't hold the tank.

The other method of getting water is by means of a pressure system, most popular in recent years. The pump puts water into a closed tank, working against air pressure, and the system can be set to deliver any degree of pressure. Both systems have their good and bad points. Under the gravity system the water does not stay quite so fresh, and in summer is apt to be warmer, being up under the roof. Nor will it give much force for a shower, lawn sprinkling, or operating some types of laundry equipment. If the system is not automatic, it requires hand starting and stopping. On the other hand, there is usually a tankful of water to tide you over for an hour or two, if the electric line suddenly decides to quit, as it occasionally does in the country. The pressure system delivers water with a constant force, and it is usually cooler and fresher, but if the current goes off, you are out of water almost immediately, which can be awkward.

For both pressure and gravity systems you should consider a reservoir as an additional source of water, especially for fire protection. If there is a hill close by the farm buildings, the reservoir problem is simple, and by merely turning a valve you can maintain running water for house and barn if the current goes off for long periods. If there is no available hill, a reservoir for fire protection is still money and time well invested. Remember that a fire engine pumps 1,000 gallons a minute. (The new "fog" type nozzles make one gallon of water do the work of ten, but not many rural fire departments are equipped with them.) Even rural-type fire engines seldom carry a tank with more than 750 gallons of water, so their efforts are usually confined to spraying other buildings than the ones which are burning, to keep the fire from spreading. A bucket brigade of milk cans rushed by neighbors in trucks will barely furnish enough water to wet down the hottest walls. With a reservoir of 3,000 to 10,000 gallons, however, the firemen may actually be able to extinguish a fire in the house, rather than watching it burn to the ground. The reservoir should be placed halfway between house and barn, and should have a manhole cover easily removed—except by children.

Older farms are usually equipped with a cistern, tapped by a hand pump at the kitchen sink. If the cistern is in good condition it can continue to serve as a welcome source of soft water, always a delight to the ladies, and as an extra reservoir in case of fire. You will probably want to pump it dry, and then get in there and clean it out before using the water. Rural hardware stores and the mail order houses have special filters which can be attached to the intake pipes of cisterns, so that you don't need to put up with the brownish, smelly water that characterizes most cisterns. If you plan to use the cistern, check the rain troughs at the eaves of the house to see that they are clean and watertight. The downspout to the cistern should have a diversion pipe to lead off the first flush of water, which is likely to be dirty. Whether the cistern is outside or inside of the house, and is used or idle, make sure it has a rat- and child-proof covering.

Remember, if you live in northern climates, to put your pipes below frost, or you may be without water, to say nothing of digging up a collection of burst pipes.

In the last few years falling levels of water tables in many parts of the country have completely changed the picture of farm water supplies. Wells that were scanty or occasionally stopped flowing in dry periods are now dry the year round. Driving a well deeper is only a temporary answer. Every farmer should be actively promoting in his area a conservation program of restoring marshes, foresting slopes and bottom lands, plugging up drainage ditches, and the like, to bring that water level back up.

Depending somewhat on the lay of the land, you should also think about the possibility of a fishpond. If there is a small stream on the property, a dam can be made, and the pond shaped, by a bulldozer. If there is no stream, but your water supply is ample, you may be able to swing a combination of garden irrigation ditch, with the overflow going to form a pond. Fertilized, and stocked with fish, a pond is supposed to equal any other acre in production of food. Your county agent can obtain plans for construction of a pond.

Even if you can't swing a fishpond, you can easily build a swimming pool for the children, which will make you the envy of the neighborhood, as well as filling in some of the gaps in recreation and sociability which children miss in the country unless a special effort is made for them. Cement manufacturers have elaborate booklets giving directions for making a pool out of concrete, but you can do pretty well by scooping out dirt, covering the bottom and sides with tar paper, and turning on the hose, at a total cost of a few dollars, instead of a few hundred or thousands of dollars. Don't forget the little matter of draining it. If it is on a slope or a hilltop, you can probably drain with a hose as siphon.

Getting running water into the house is only half, or maybe less than half, of the story. Getting it out again is just as important, which means a sewerage system. If the house is on high ground, with a slope away in any direction, the problem is simple. You can build a septic tank and dry well, and provide an overflow which will take care of any excess effluent. However, many a farmhouse is located in a pocket, or on level land with clay undersoil, which simply won't carry away the 200 gallons of water which a medium-sized family will use in a day. You can get rid of the excess effluent in the dry well by constructing what is known as a "dry field"—long runs of drainage tiles near the surface of the ground—or you can pump out the effluent every two or three weeks. One way will give you a lot of work and some expense at the beginning. The other will give you a little work, and almost no expense, at intervals. If you spray it on the land—or even the garden—you will be amazed at the lush growth produced. The only trouble with pumping is that the dry well always seems to need it just when you have visitors from town, and the wind is in the wrong direction.

Whatever system you use, any plumber who sells you the equipment, as well as the agricultural colleges and the county agents, have complete plans for putting in a sewerage system. I put in one myself on my first farm, with a week's digging in spare time.

But before you spend too much time and money fixing up the old farmhouse, better consider whether you want to live in it at all. If your farm is large enough to require a married couple for help, you are likely to be better off giving the old farmhouse to the tenant or sharecropper, and building a new one for your own living. One kitchen and one living room can become pretty congested with two families using them. There is also the consideration of different hours of work for master and man. If you run a dairy farm, for instance, and leave most of the milking chores to the tenant, he will be pounding in and out at hours when you want to sleep, and he will constantly be filling the house with the rich odors of the barn. Above all else is the consideration that if you are moving to the country in order to live your own life, you can't really enjoy it if you have someone else under foot all the time.

Your long-range plans for the other farm buildings depend, naturally, on what type of farming you plan to pursue, and what farming methods you intend to follow. The mistake many beginning farmers make is in starting at once to tear down buildings they consider unnecessary. At present costs of materials and labor, even the poorest building can usually be converted profitably to other uses, without running into the expense of new construction. For instance, an old machine shed or horse barn can easily be changed into a loose barn for cattle, or a shelter for sheep. If the barn is equipped with stanchions for dairy cattle, and you want to specialize in beef cattle, the stanchions can be knocked out in an afternoon with a maul, and gravel spread to make a level floor. It is harder, however, to convert a plain barn into a dairy barn by adding stanchions, for that means feed mangers in front and gutters behind, plus a driveway to haul out the manure. In such a situation, the barn might well remain as a loose barn, and an adjoining small building be built or converted into a milking parlor. In the same way, hoghouses can often be converted into an abode for chickens, and vice versa.

Type of farm operation also plays a part in determining what to do with the buildings. If you plan to change, for instance, from making loose hay and storing it in the haymow, to production of chopped or baled hay, which take up only half the space, you may need to strengthen the floor supports of the haymow to prevent a collapse with the doubled weight. In that case you might take out half of the haymow floor, use part of it to strengthen the remaining half, and the rest to build self-feeders around the sides of the open half. In that way the greatest load of hay would be resting directly on the ground, and you would save yourself a lot of time in pitching down feed.

A concrete-floored barnyard is another long-range possibility. Many farmers are now installing them, to keep the animals cleaner, and to save the manure which is otherwise largely wasted.

Turning from buildings to fields, the future development aspects usually take more time, but can be equally interesting and rewarding. If the fencing is old and poor, you may want to seize the opportunity, when building new fencing, of rearranging the layout of fields. If possible, you want a central lane, accessible from both the farmyard and the barnyard, to reach all the fields with the shortest possible travel time. This will save you time in hauling out manure, and moving the livestock from barn to field and back again. It will also save you tillable land and fencing, for lanes are strictly unproductive.

For flatlands, windbreaks and shelter belts are possibilities to consider. For rolling lands, you may have slopes which will yield a bigger crop if planted to quick-growing trees for pulpwood, nut trees for harvest, an orchard, or Christmas trees. On a poor, sandy hillside, for instance, too wretched to use for pasture or crops, you should be able to grow a fine crop of Christmas trees in less than ten years.

In fact, all the land too steep for the plow or too thin for a field crop should be turned into woods as soon as possible. With land that thin, don't try to make it produce a double crop by turning in any cattle. They won't get anything of value to eat, but they will infallibly destroy all the new seedlings, and provoke erosion with their trails. (See Chap. 18, "Orchard, Woodlot and Fencerows.")

Whether management of the remaining fields to prevent soil erosion and conserve moisture should be part of any long-range plan is open to debate. If erosion is started and is progressing fast, you'll have to do something quick, rather than at some future day, or you won't have any farm left. Nevertheless, if the financial pinch of starting out is too acute, soil-saving that involves an outlay for bulldozing terraces may have to wait until cash is available. Meanwhile the field should be put into a heavy cover crop and left for hay and pasture, until money is available to make it suitable for rotation crops.

Whatever the long-range plan, let it also include some elements of pure pleasure, to make the farming experience livable. Some of these might be a hedge of native shrubs and flowering bushes, grubbed from the woodlot and placed near the house for their beauty and interest; a fencerow left to run wild as a shelter for small animals that the children can hunt or merely observe; good cover and water on some unused portion of the farm for game birds; even a rough wall of field stones to idle against in the first warm days of spring and the chilly ones of fall.

ADDITIONAL READING MATERIAL: *Man Around the House*, by Robert Engels (Prentice-Hall); *Concrete Handbook of Permanent Farm Construction* (71 pp., ill.), Portland Cement association, 33 W. Grand Ave., Chicago; *Farm Pond Building*, J. R. Haswell and C. G. Burress, C320, Pennsylvania State College, State College, Penn.; *Developing the Farmstead*, by R. B. Hull, B 345, Purdue University Agriculture Extension Service.

Free from U.S. Dept. of Agriculture: *Stock-water Developments, Wells, Springs, Ponds*, F 1859; *Sewage and Garbage Disposal on the Farm*, F 1950; *Making Cellars Dry*, FB 1572.

From Supt. of Documents: *Use of Concrete on the Farm*, 10¢, FB 1772; *Sewage and Garbage Disposal on the Farm* (including information on septic tank systems), 10¢, Cat. No. A 1.9:1950; *Fire Safeguards for the Farm*, 10¢, Cat. No. A 1.9:1643; *Simple Plumbing Repairs in the Home*, 5¢, FB 1460; *Protection of Buildings and Farm Property from Lightning*, 10¢, Cat. No. A 1.9:1512; *Farm Fishponds for Food and Good Land Use*, 10¢, Cat. No. A 1.9:1983; *Fireplaces and Chimneys*, 15¢, Cat. No. A 1.9:1889; *Landscaping the Farmstead* (88 pp., ill.), 25¢, No. I 16.54/3:189; *Plans of Farm Buildings—Northeastern States*, 60¢, Cat. No. A 1.38:278; *Southern States*, 60¢, Cat. No. A 1.38:360; *Western States*, 60¢, Cat. No. A 1.38:319.

24

TOOLS FOR THE JOB

If you are no good at carpentry or metal working, don't buy tools for such jobs—hire it done.

For the one-acre or subsistence farm, about $50 worth of small hand tools should be enough.

For five to ten acres, add a car trailer, and either a garden tractor or a small, secondhand regular tractor, depending on your inclination. They will cost about the same.

Keep away from the snare of dual-purpose tools. They are not worth the extra exertion and cost.

For larger farms, consider the possibility of custom work, rather than investing in thousands of dollars worth of machinery.

Hay-making and harvesting equipment will depend on the type of farming you have determined to follow.

The machinery for one method will usually not work for another.

THE tools and machinery you can use on any-size farm depend on your own skill and ability perhaps more than on the necessities of the work. If you are good at carpentry, you probably already have a full set of tools for all tasks. If you aren't any good at it, why bother with more than a saw, hammer, chisel and screwdriver, plus a brace and bit? In the same way, a good machinist will have plenty of metal-working equipment, and know what to do with it, while the duffer, with wrench and hammer, can disconnect the broken piece, take it to the local blacksmith shop, and get an equally good job done. Each will probably put in about the same amount of time, but one will pay out a little money besides—which may, in the long run, represent less than the investment in a full set of tools.

Nevertheless, even if you do yourself the things you are handy at, and hire the others done, you will need a certain bare minimum of tools and machinery, no matter what the size of the farm. You can add onto the list at any point, but don't just be tempted by the shiny look of new things in the store. Every new tool costs money, takes storage space and care, and if rarely used, actually represents a loss.

First of all, for the smallest farms—even a subsistence acre—you will need: a long-handled shovel, a pick or crowbar, hoe, rake, wheelbarrow, pipe and monkey wrenches, hammer, saw, screwdriver, chisel and brace and bit—about $50, more or less. From the point of view of the man who steers it, one of the greatest modern inventions is the pneumatic rubber tire for wheelbarrows, and they are well worth the added cost of about $12. You can haul heavier loads, over rougher ground, and with less effort, than you can with the old iron-tired wheelbarrow. A truck gardener's wheelbarrow has flat bed and removable sides, for easy handling of flats and boxes. If you are not going into truck gardening, get one with steel sides.

If you have a few more acres, say up to five or ten, the most important addition to the above would be a two-wheeled trailer for the car, or a jeep. The trailer can be one of those fancy steel box affairs, beautifully balanced, with a stake rack for hauling livestock—or you can pick one up secondhand for anywhere from $10 to $50, make your own rack if you need one, and probably get on just as well.

Whether you actually can use a two-wheeled power garden tractor depends somewhat on your own inclinations and skill. Even the best are subject to occasional repair and temperamental fits, and most are not heavy enough for real farm jobs like plowing. Some subsistence and "acre farmers" swear by them, while others swear at them. One good way to find out whether you can really use one is to borrow a garden tractor from a neighbor, and see how you like it.

County agricultural agents and instructors I have talked to are inclined to sniff at garden tractors as impractical. They say you can do the same job better and cheaper with a small secondhand tractor —as well as doing other work that the garden tractor is incapable of performing, such as plowing and hauling heavy loads. For example, with a tractor you can make and haul in hay, haul out manure, plow and disk, saw wood, run various grinders and power tools, and use it to start your car in cold weather—assuming you can get the tractor to start.

Perhaps this is as good a place as any to say a word about—or against—dual-purpose tools. They have a persuasive plausibility about them: two or more tools on the same handle, and for little more than the price of one tool. But it is the nature of such beasts to be not very good for either of the dual purposes, and while you are using one facet of the dual tool, you are carrying the dead weight of the other part. During a full day's work, this can amount to a considerable expenditure of manpower—maybe enough to do a couple of extra rows of beans before nightfall.

While it may hardly seem necessary to mention it, a complete set of farm clothes is also necessary. You need heavy work shoes, shoe-top overshoes, or higher, for winter and wet weather, overalls, and windproof jacket. Canvas gloves which can be washed, and thrown away when worn out, are about the most practical, except for severe winter conditions. If you don't like the sight of an occasional hayseed in your cap, you can invest in one of those patent-leather caps with visors, and look like the old country. Have enough extra pairs of overalls so that there is always a clean pair to "go to town." A good farmer takes pride in looking clean, even though he may be in work clothes.

The equipment for the small-sized farm, and anything on up from there, depends largely on the pattern of farming you have chosen to follow. Before investing in a lot of machinery, the novice would do well to consider custom work for most of his farm tasks, at least until he has had time to get his feet firmly planted on the land, and knows what machinery he actually needs. Custom work is now fairly general throughout the country, in many areas replacing the old co-operative arrangements in which several farmers banded together for a common task, with commonly owned machinery.

Thus the old "threshing ring" of half a dozen farmers jointly owning a threshing machine and working at each other's places until all the grain was in, is being replaced by the neighbor with a combine, who does the job of threshing all the standing grain, at so much per acre. After the combine has clipped the field, a side-wheel rake gathers the straw in windrows, and another neighbor, or perhaps the same one, comes in with a baler, and neatly packages the straw for stacking in the barn. Considering the cost and life of machinery, and the expense of storing it under shelter, a man can do about as well or better with custom work than by owning his own machinery. The one great advantage of having your own machinery is being able to do the work at just the right time for a maximum crop. Failure of the baler to arrive at the proper time when you have a lot of hay down, for instance, could be ruinous to the crop, and hence, in custom work, it is wise to deal with trustworthy individuals. Aside from keeping the machinery investment to a minimum, custom work has certain other advantages. If some part of the machinery breaks, the owner of the machine repairs it at his cost and time, not yours. Since he is a contract worker, your liability in case of accident is virtually nil, and you are not yourself handling dangerous machinery.

Suppose, though, that custom work is not easily available, or you just want to do the job yourself. Unless bought cheaply at second-hand, which takes a shrewd eye and experienced judgment of condition, you can expect to lay out several thousand dollars for machinery, regardless of the type of farming you have chosen.

The biggest single expenditure is for a tractor, needed on virtually every type of farm. For the small place, a small tractor, capable of pulling a single-bottom plow, light manure spreader, small disk, and the like. The larger the farm, generally, the larger the tractor. Many an American farm still carries one or more teams of horses, but it is sentiment, rather than economy, which dictates their use. They are insatiable eaters, require extra-good fencing, and a lot of chore time. If you want to keep one or more horses for reasons of affection, that of course is your privilege, but it should be done with full recognition that they won't do as much work, or as cheaply, as a tractor. It may seem harsh thus to dismiss man's noblest friend to the glue factory or the pleasure-riding class, but we are concerned here with economical and efficient methods of farming, rather than with sentiment.

One of the major disadvantages of trying to combine horses and tractors is that they require two complete sets of hitches for each machine—a short one for the tractor, and a long tongue for the horses. The changeover from one type of power to the other can take hours, and you can be sure that if you want to do a heavy job with the mower and tractor, the horse hitch will be the one that was last left on the mower.

Whatever the crop, you are going to have to fit the land first, and this is true even if you plan to keep the whole farm in grasses and legumes most of the time.

The basic tool is the moldboard plow, pulled by horse or tractor. For heavy sod, or where there is a coating of mulch or vegetation above the soil, a colter (free-running sharp wheel) just ahead of the plow, is usually necessary to do a clean job of plowing.

After the land is plowed and has dried out a bit, it is worked with a harrow, disk, or spring-tooth quack digger, depending on conditions and the type of seedbed required. This is the time, also, when the farmer on rocky or glacial soil brings out the stoneboat and hauls off the rocks heaved up by frost and plow.

Among newer devices on the market is the "subsoil plow," which contains an added steel blade to stir up soil packed by years of pressure from the sole of the regular plow. There are also machines which plow, harrow and mix the soil in one operation. They do not cover as much ground in one day as the standard machines, and they are relatively costly. Use of these new devices depends partly on the inclination of the farmer, and partly on specialized local conditions.

At the other extreme are loose soils which need to be firmed to produce a good seedbed. Old-time farmers used to have a section of tree trunk with an axle through the middle, which they hauled across the field for this purpose. The modern equivalent is a corrugated metal drum pulled behind the seeder, and filled to the desired level with water.

Among the most modern devices are chemical weeders: long booms of pipe, swung from the sides of the tractor, through which chemical weed killers are forced by a pump attached to the power take-off of the tractor.

Because farming methods and crops differ throughout the country, no single outline of machinery can be set up to cover all types of farms, but you can choose from among the following, according to your own pattern of farming:

Except in the extreme South, haymaking is virtually standard practice on all farms, but the methods differ widely. You will have to determine ahead of time what type of haymaking you choose to follow, and then pick machinery which will do that kind of job.

The simplest and most inexpensive tools for haymaking are a mower and buck rake, assuming that you want to go beyond the scythe and the old wooden hand rake of our ancestors. You can "knock down" the standing hay with a regular sickle-bar mower pulled by tractor or horses, or mow it down yourself with a garden tractor. The advantage of the buck rake is that the hay does not need to be raked into windrows first. The rake itself is composed of eight-foot limber poles, about six inches apart, and tapered upward at the open end. It can be attached to the front of a tractor or even a pickup truck. The rake is pushed along under the hay until a load is obtained, and then, by means of cable or rope, pulled to near vertical, pinning the hay between the rake and upright bars at the front of the vehicle. To unload, drop the rake. It has the advantage of shattering few leaves, because the hay is handled only once.

After the buck rake, the next step toward more complicated

machinery is a mower and either a dump rake or side-delivery rake. The dump rake gathers the hay into more or less even windrows, depending on how expert the raker is in depressing the dump pedal. The side-delivery rake uses steel fingers powered by a gear on the wheel, to toss the hay into one long windrow which coils around the field like a giant snake. For small fields, a pitchfork can collect the windrowed hay into cocks, ready to be pitched up on a wagon.

For larger fields, several methods can be used. The most popular, up to a few years ago, was the hay loader pulled back of a wagon, which straddles the windrow and delivers the hay at the back of the wagon, from where it is forked, with considerable exertion, to form a compact hayload.

Newer methods are the hay baler and the forage harvester. Each straddles the windrow, as does the old-style hay loader. The harvester chops the hay into pieces a few inches long and blows them into a big wagon box. The wagons are hauled to the barn, where another blower whisks the hay into the mow. The baler either drops the bales behind it on the ground, from where they are tossed later onto wagons, or else feeds directly into an attached wagon.

When green hay is being put directly into the silo as "pickled pasture," you need a mower and rake, and either a forage harvester to chop it up, or hay loader to pile it on wagons, to be hauled to a chopper at the silo.

The newer methods of hay and corn harvesting, in which the hay, or the corn for silage is chopped up into small pieces by a forage harvester, also require a new type of wagon. Usually these are made to order by the local blacksmith, or you can make them yourself. For fairly level fields, where there is little danger of tilting, they are often made on a two-wheel base, with a platform six to eight feet wide, and from fourteen to eighteen feet long. The sides usually rise about six feet, of solid wood, and some farmers add another two feet of wire netting in a wooden frame. Many are now self-unloading.

Where hauling is more precarious, some farmers use four wheels to support the platform, which requires less maneuvering and heavy lifting, in shifting the trailer on and off the tractor hitch. Builders usually pitch the bottom of the trailer toward the center, so that most of the unloading is done by gravity, through a small door at the bottom center of one side. Hitched directly to the tractor which pulls the forage harvester, these big boxes on wheels permit one man to work alone in harvesting, instead of the former requirement of one person to drive and another to load.

To put the hayseed into fitted ground you can get along without any machinery at all, though this is wasteful of expensive seed. The seed can be broadcast by hand from a sack slung from the shoulder, as our forefathers did. Unless you are an old hand at it, however, the seeding is likely to be uneven and costly. The next step up in hand operation is a "fiddler," a bag with wooden bottom which drops the seed in a trickle onto a grooved platform whirled with a crank. Still more mechanical is a hayseeding attachment on a standard grain seeder, which scatters the hayseed at the same time the "nurse" crop of grain is sown.

All the small grains, such as oats, barley, rye and wheat, can be broadcast by hand, or by a fiddler, but are most easily sown with a regular seeder, which is a long box between wheels, feeding down by pipes into drills which scuff into the soil. (All of the small grains, particularly oats and rye, can be cut while still green and used for hay.)

When harvested as grain they can be cut with a mower, stacked, and pitchforked into a threshing machine or combine, but the easier method is to use a binder or combine. The binder cuts the grain, ties it into bundles, and drops them off in the field. Here they are set up into irregular shocks by hand, to dry out. In the days of the old steam threshers, the shocked grain was later hauled to the barnyard and stacked into meticulously shaped piles which were the pride of each farmer. In recent years, however, in the small threshing rings, half a dozen farmers gather with wagons, and threshing is done either in the field or the farmyard, depending on where the farmer wants his strawstack. Combines are also used to thresh directly in the field, and the straw is either dropped as mulch or later stacked or baled.

Corn, the other major farm crop, is usually planted with a two-row planter. If it is intended for silage, the farmer usually "drills" it in, letting a kernel drop about every eight inches or a foot. That way he won't be able to cultivate out the weeds within the row, but they will merely add to the silage. Where ear corn is desired, or weeds are terrific, the farmer plants in "check rows." A long wire, with knobs at intervals corresponding to the width of the rows, is set up from one end of the field to the other, and as it passes through the planter mechanism, drops the corn in hills. This permits the farmer to cross-cultivate his field. Usually, when the corn is planted in hills, it is also given a shot of commercial fertilizer a couple of inches from the hill, to give it a quick start.

The corn is cultivated with a harness of hoes slung over the tractor. At harvest time, if it is going into the silo, it is cut with a corn binder into bundles for the corn chopper at the silo, or a forage harvester can chop it, ready for the silo. If ear corn is desired, it can be cut with a binder and the bundles gathered into shocks to dry out. This method is used when the corn is to be shredded later at the barn, the ear corn going into the crib, and the shredded stalks blown into the barn for bedding and extra feed.

Sometimes pigs, beef cattle, or turkeys are turned directly into the standing corn to do their own harvesting. Another alternative is to hire a corn picker, which husks the ears and spouts them into a wagon behind, leaving the stalks in the field. A more recent machine is the "Cornbine," which husks the ears and shreds the stalks.

MARKETING, CO-OPERATIVES, FARM ACCOUNTS AND TAXES

Marketing is where the farmer can lose his shirt.

The greatest profit is in direct sale to the consumer—and also the most work.

In selling first-class breeding stock, the sky and the buyer's purse are the limit.

The simplest way of selling livestock does not bring in the most money.

Keep "cow jockeys" out of your barn.

If you can store a field crop, your price will be higher.

But selling hay direct from the field saves labor.

Farm accounts are an inventory, like your soil testing.

To claim income tax deductions, you must have the records.

"Loving up" the local assessor of real estate.

If your total taxes are near those of city property of equal sale value, something is radically wrong.

Consolidated schools can sometimes wreck the farm tax picture.

Drainage districts—where your money floats away.

THE many skills which together make up a good farmer should include astuteness in selling his crops and livestock, as well as in the technical and practical knowledge, to say nothing of the energy that he has expended in producing them. The history of early wheat farmers on the Great Plains, the plight of cotton sharecroppers, and the depression prices paid for milk and hogs point up the sad truth that a man may be good at raising crops and still lose his shirt.

There is no room here to go into the vast complexities of farm marketing problems. What will be attempted is to sketch the common sales avenues, with some indication of their pitfalls and benefits. The local situation will govern many sales—but such local situations can be changed by determined individuals and groups. Make sure it is not the other side which is doing the changing.

Since the government has entered the picture, to support a price for farm products which bears a relation to the things farmers must buy, the farmer is no longer compelled to accept whatever price is offered in the market place. (I am not attempting here either to defend or attack the federal system of price supports, but merely to point out that they exist, and in many cases exercise a dominant role in the market.)

You can sell direct to the consumer, to a broker or processor, or through a marketing co-operative, depending somewhat on what facilities exist, and where you can get the top dollar. The sale of some products, such as raw milk, is strictly regulated, but in general the field, meat and vegetable crops are yours to dispose of in any way you see fit, to any buyer.

By and large, you will get the most profit by selling direct to the consumer, since you eliminate the middleman's profit and the handling costs. Usually you will spend a little more of your own time in such transactions than if you sold to a broker or through a co-operative. An egg route in town is an example of high-profit direct sales, because you get the top retail price, or better, for strictly fresh eggs. Garden produce, small fruits, eggs and dressed chickens, carried to the door or sold from a roadside stand, are the chief products from which you can extract the full retail price. A roadside stand, incidentally, is not much good unless you are on a busy highway.

Livestock sales fall into two categories. If you are selling breeding stock, the quality of your animals and the extent of the buyer's purse set the limits. (See Chap. 5, "Plan to Be a Specialist.") Exceptional breeding stock will fetch extraordinary prices, but ordinary animals will probably bring not much more than the market price for the carcass. If you happen to be raising a purebred strain of sheep, for instance, you can often dispose of breeding stock by small ads in breeder's magazines, and get fancy prices for your animals. Because of the value of human relations, sales to your neighbors are likely to be at a slightly lower price than to strangers. Sometimes this is good business.

In selling livestock to be processed into meat the farmer bumps into a more competitive area. The simplest way is to hire a trucker to take the animals to a packer, or commission agent, and then take the packer's word for the grade and quality of the animals. Since all the big meat packers are about equally efficient, and sell their cuts on a highly competitive retail market, the packer's only chance for a "spread" comes in astute buying, either at the plant or elsewhere. Many packers now have field agents who will visit your farm on request, and grade your animals before you sell them, so that you can determine just what price you will receive—and whether you can do better in some other place.

An alternative to the packer is the co-operative livestock marketing or shipping association. Here the farmer, for a small percentage of the sale price, consigns his livestock to the co-operative, which has trained sellers at the big markets. The co-op seller is qualified to argue with the packinghouse buyer as to the proper grading of the animal, and can if advisable hold the animal for another day, or ship it to another market, if that seems wise, instead of dumping it at whatever price is offered, as a trucker would do. The growth of such livestock marketing co-operatives is a sign that farmers believe they obtain a better price through that means. The co-ops possibly also have an influence on the general market through the competition they offer to packing house buyers.

Some farmers sell direct to "cow jockeys"—individuals who drive up to the farm in a truck, and offer to pay spot cash for old, crippled, or unwanted animals. Sometimes the payment is by check, and sometimes the check comes back, but the cow doesn't. Since both the farmer and the cow jockey are guessing at the weight of the animal, and the cow jockey is a shrewd buyer or he wouldn't stay in business a week, it is easy to see who is going to make money on the deal. The only convenience of a cow jockey is that he saves a trucking fee and brings in immediate cash. Many farmers, however, will not allow a cow jockey in the barn, or even on the property, because of the danger of transmitting bovine diseases.

Other sales outlets for old, crippled or diseased animals are to places like fox farms and dog food makers. Dead animals can usually be disposed of to a rendering plant for a few dollars, and the rendering plant will do the hauling away. Papers with rural circulation usually carry notices of rendering plants in the want ads.

Marketing field crops, such as corn, oats, rye and the like, depends somewhat on your storage facilities. The price is usually lowest just at harvest time, and increases gradually until the next harvest. If you have storage space which will keep the crop from spoiling, you can usually get a better price by waiting. Here again local feed mills and the co-operative marketing associations compete for the farmer's business. The co-op price may be a little higher, but you will wait longer for your money. Many marketing co-ops make an immediate partial payment, which is somewhat below the expected low point of the market, and then send the balance later, after the co-op's total transactions in that crop are completed.

Some crops can be sold to advantage directly from the field, even in advance of full maturity. For instance, baled hay can be stored in the barn, and sold later in the winter at considerably higher prices, but part of this higher price will be absorbed by a 20 per cent loss in weight. There is also the convenience of having the buyer pick up directly from the field as the hay is baled, sparing you the work of hauling it to the barn, stacking it, and waiting for a sale later.

Many specialized crops, such as fruit, are sold on a contract basis to commission firms before they are ripe. Your neighbors can advise you on reliable dealers and common prices.

As you sell, the measure of your success in the farm enterprise will show up in your accounts, which also prove valuable when income tax return time rolls around. Just as a soil test is basic to the proper handling of field crops—like the merchant's inventory, it tells you what goods are on hand, and what you need to keep in business—adequate records will tell you whether you are standing still, gaining, or falling behind.

The records need not be elaborate. The farm implement companies, the dime stores and the state agricultural colleges all have simple record forms available, all free except for the dime store. On these record sheets you put down, in the proper columns, the amounts you pay out for feed, stock replacements, machinery, taxes, interest, fertilizer, and the like, and the proceeds of your sales. Between the two, you can get a good idea of how you are doing.

Farmers can claim many deductions on federal and state income taxes for expenses of operation, but you will need these simple records in order to substantiate your claims. A dairy cow used for income purposes, for instance, can be depreciated just like a machine in a factory, and her original cost, less the salvage when she is sold, deducted as a business expense item. The same is true of other capital expense outlays.

In addition to federal and state income taxes, your farm, regardless of its size, will also be subject to the ministrations of the local real estate assessor. He is likely to be a man from the neighborhood, elected by the township to do a part-time job for a brief period in the spring. Often the local assessors are given a one- or two-day course of training by the state or county, before they begin their work. Obviously a neighboring farmer, cast in the temporary role of an assessor, is no skilled auditor, but neither is he the village idiot. He is a man much like yourself, who wants to put a fair value on the various classes of land, the buildings, and the personal property like livestock and machinery. If your own attitude is that of wanting to pay your fair share of the cost of local government—but no more than your fair share—you ought to get on fine with the assessor.

It is worth your while, in friendly discussion of land values, to take time to check his book to note what values he has placed on neighboring properties, just to make sure that you do not suffer the occasional fate of a newcomer in being loaded with an extra-high assessment. An easy way to do this is to suggest that you are naturally interested in seeing that you are paying your fair share, and therefore would like to see how your assessment stacks up with that of neighboring similar properties. After all, the records are public. If you are dissatisfied with the assessment, there is usually an appeal body which meets later, before the tax roll is completed, and the clerk of the municipality can tell you when this organization holds its sessions.

The taxes you pay are based on two factors: the assessed value of your property, and the tax rate or levy. Thus you might have a low assessed valuation, and a high tax rate, or a high valuation and low rate, and come out about the same. Often in the country property is assessed at considerably below its normal sale value, and the taxes you pay, even though the rate is high, are correspondingly low.

Regardless of the assessment or rate, the total real estate taxes you pay on country property should be considerably lower than the tax on property of similar sale value in the city or in urban districts. If they are nearly the same, something is radically wrong, because usually the amount of municipal services furnished in the country is far less than that in the city. To name but a few, you are doing without garbage and trash collection, street lights, adequate fire and police protection, library, and usually without the extra qualities and expense of a big city school.

A school, however, may be the principal reason for a high total tax. Some of the new consolidated school districts in rural areas have overcome an educational deficiency of scores of years standing, by erecting large-scale new schoolhouses with bonds to be paid off in as little as fifteen years. There are instances in my own state where the school tax alone in a consolidated district is greater than the annual rental value of the farm. In such a case you are obviously operating an educational charity, rather than a farm business enterprise.

While it is true that the farm owner may have benefited in years past from low school taxes as a result of low educational opportunities, nevertheless it comes as a jolt to the new owner to be asked to make up this deficiency in just a few years. The existence of a low school tax, for instance, may have played a part in causing a higher purchase price for the farm, since the rental net income, and thus the capitalized value, would have been higher with a low tax.

In addition to checking before farm purchase on the existence or prospect of an expensive new consolidated school, be on the lookout also for drainage districts and their unexpected special taxes. Sometimes a whole valley and the farms on the ridges draining into it are included in a special drainage district, which may be engaged in costly operations ruinous to your tax picture. The theory of drainage districts is that the rainfall draining off your ridge farm is causing a flood in the valley—and possibly it is. But unless your farm is in the flatlands, you will not get one penny of benefit from a drainage ditch, unless it happens to keep your access road clear of water. It is the farmer in the low spots, whose farm purchase price should have reflected the excess water danger, who is practically the sole beneficiary of a drainage district. The low-spot farmers, being the ones most vitally interested, are also most likely to be the members of the drainage district commission, and hence the persons most interested in spending, rather than saving money.

ADDITIONAL READING MATERIAL: From U.S. Dept. of Agriculture: *Useful Records for Family Farms*, F 1962.

From Supt. of Documents: *Useful Records for Family Farms*, 10¢, Cat. No. A 1.9:1962; *Farm Family Account Book* (with ruled forms), 35¢, Cat. No. A 42.2:F22/2.

Appendix of Useful Information and Tables

Breeding and Gestation Table

	Can Be Bred After	In Heat	Heat Cycle	In Heat After Birth	Time to Breed After Birth	Gestation Period
Mare	24-36 mo.	3-7 days	3 wks.	3-17 days	9th day	340 days (approx. 11 mo.)
Cow	15-18 mo.	3 to 48 hrs.	3 wks.	28 days	6-8 days	283 days (approx. 9 mo.)
Sow	6 mo.	1-5 days	3 wks.	3-9 days	8½ wks.	114 days
Ewe	6 mo.	1-3 days	13-19 days	6-7 mo.	summer and fall	150 days (5 mo.)

Average Daily Livestock Consumption of Water

Each Milk Cow	25 gal.
Steer or Dry Cow	12 gal.
Horse	12 gal.
Sheep	1½ gal.
Hog	2 gal.
100 Chickens	4 gal.

Weight of Grain and Produce Per Bushel, in Pounds

Alfalfa seed 60
Apples (green) 48
Apples (dried) 25
Barley 48
Beans (white) 60
Beans (lima) 56
Bermuda grass 35
Bluegrass seed 14
Bran 20
Brome grass seed 30
Buckwheat 50
Carrots (without tops) 50
Cherries (with stem) 56
Cherries (without stem) 64
Clover seed 60
Corn (shelled) 56
Corn (ear, husked) 70
Corn (green, sweet) 35
Cotton seed 33
Cucumbers 48
Eggplant 33
Flaxseed 56
Grapes 48
Hempseed 44
Lime 80
Malt 44
Millet 50
Mustard seed 58-60
Oats 32

Onions 57
Onion sets 28-32
Orchard grass seed 14
Peaches 48
Peas (dry) 60
Peas (stock and
    green) 30
Peppers 25
Popcorn (ear) 70
Popcorn (shelled) 56
Potatoes 60
Rapeseed 50-60
Redtop seed 14
Rutabagas 56
Rye 56
Salt 50
Soybeans 60
Sorghum or cane
    seed 50
Spinach 18
Sudan grass seed 40
Sunflower seed 22
Sweetclover 60
Timothy seed 45
Tomatoes 53
Turnips 54
Walnuts 50
Wheat 60

(These figures are also intended for selling by weight. Different varieties of grains and vegetables will show more or less weight per bushel by volume, as well as differences due to degree of moisture.)

Quantity of Seed Per Acre

(In pounds for all but onion sets and potatoes, which are in bushels)

Alfalfa	8-15	Oats	64-96
Brome grass	8-12	Onion sets	10-12 (bu.)
Buckwheat	25-50	Orchard grass	15-25
Bluegrass	25-50	Potatoes	10-15 (bu.)
Barley	95-100	Redtop, in chaff	25-35
Corn	8-11	Redtop, fancy	6-8
Clover, red	10-15	Rye	70-90
Clover, white	4-6	Stock peas	50-80
Clover, alsike	4-6	Sunflower	8-10
Flax	55-80	Timothy	8-12
Hemp	40-60	Wheat	75-110
Millet	35-40		

(In seeding mixtures of grasses and legumes for hay or pasture, such as alfalfa and brome grass, or clover-timothy-alsike, figure a total of about 15-18 pounds of seed per acre, dividing this total up among the combination of seeds.)

To Measure Ear Corn in Crib

Find number of cubic feet (length times width times height) and multiply by 4, then divide by 10, to get number of bushels (about 2½ cubic feet per bushel). If corn is well filled, dry and well settled, divide by 9 instead of 10. If damp and not well filled, divide by 11.

Grain in Granaries and Wagon Beds

Multiply number of cubic feet by .8036 to find number of bushels. Or find cubic feet and deduct one-fifth.

To Measure Hay in Mow in Tons

Find cubic contents by multiplying height, width and length. Divide by 350 for old, settled hay; by 425 for hay one to two months old; and by 500 for hay less than 20 days old.

To Measure Tons of Hay in Stack

Multiply the width by the overthrow (distance from the ground on one side over the top of stack to the ground on the other side at an average height) and divide result by 4. Multiply the result of this division by the length, to obtain approximate cubic contents. Divide by 500 for hay in stack one to two months; divide by 400 for hay in stack over two months.

Capacity of Silos

One ton of silage occupies 50 cubic feet. Find cubic feet in silo by multiplying square of inside diameter by 0.7854, and then multiplying this result by height of silage in feet. Divide by 50 to obtain number of tons.

Capacity of Cisterns, Reservoirs, Tanks and Barrels, in Gallons

For circular receptacles, multiply square of diameter by 0.7854, and multiply this result by height. Divide by 7½ (exact figure: 7.48).

For rectangular receptacles, multiply length times width times height, and divide by 7½.

To Find Cost per Pound of Items Sold by the Ton

Multiply the number of pounds by the price per ton, point off three places and divide by 2.

Land Measure

1 rod: 5½ yards or 16½ feet
1 mile: 320 rods, or 1,760 yards, or 5,280 feet
1 acre (208 by 208 feet): 160 square rods
1 township: 36 sections, each 1 mile square
1 section: 640 acres
1 quarter section: half a mile square, or 160 acres
1 eighth section: half a mile long and a quarter mile wide, 80 acres
1 sixteenth section: a quarter mile square, 40 acres

To find number of acres in parcel of land, multiply length by width in rods and divide product by 160. When opposite sides are unequal in length, add them, and take half the sum for the mean length or width.

Cubic Measure

1 cubic foot: 1,728 cubic inches or 7½ gallons
1 cubic yard: 27 cubic feet
1 gallon: four quarts or 231 cubic inches (One gallon of water weighs 8½ pounds.)
100 pounds of milk: 46 quarts
1 barrel: 31½ gallons (There are 3.8 cubic feet of cement in a barrel, and 42 gallons of oil.)

Square Measure

144 square inches: 1 square foot
9 square feet: 1 square yard
30½ square yards, 1 square rod

Cordwood Measure

A cord of wood is 4 feet wide, 4 feet high and 8 feet long, and contains 128 cubic feet. To find number of cords in a pile of cordwood, multiply the length, width and height in feet, and divide by 128. Dry oak weighs 3,600 to 3,800 pounds per cord; green oak 4,400.

To Find Height of Building or Tree

Set up a stick vertically and measure its shadow. Measure length of shadow of building or tree. Length of shadow of building, times height of stick, divided by length of shadow of stick, equals height of building or tree.

Cement Mixtures

For small jobs:

| ⅔ pail water | 2¼ pails sand |
| 1 pail cement | 3 pails gravel |

For larger jobs, materials required per cubic yard (27 ft.) of concrete:

	Sack of Cement	Sand (cu. yd.)	Gravel (cu. yd.)	Largest Size of Gravel
Floors, steps, walks, tanks, silos, etc. (1:2¾:4 mix)	6¼	⅔	¾	1½ in.
Thick footings, foundations, etc. (1:2¾:4 mix)	5	⅔	¾	1½ in.
Thin reinforced concrete in milk tanks, fence posts, slabs 2-4 in. thick, etc. (1:2¼:2½ mix)	6½	⅔	¾	¾ in.
Very thin concrete, for lawn furniture, and slabs 1-2 in. thick (1:1¾:2¼ mix)	8	⅔	¾	⅜ in.

Index

A CATALOGUE OF SELECTED DOVER BOOKS
IN ALL FIELDS OF INTEREST

A CATALOGUE OF SELECTED DOVER BOOKS IN ALL FIELDS OF INTEREST

CELESTIAL OBJECTS FOR COMMON TELESCOPES, T. W. Webb. The most used book in amateur astronomy: inestimable aid for locating and identifying nearly 4,000 celestial objects. Edited, updated by Margaret W. Mayall. 77 illustrations. Total of 645pp. 5⅜ x 8½.
20917-2, 20918-0 Pa., Two-vol. set $8.00

HISTORICAL STUDIES IN THE LANGUAGE OF CHEMISTRY, M. P. Crosland. The important part language has played in the development of chemistry from the symbolism of alchemy to the adoption of systematic nomenclature in 1892. ". . . wholeheartedly recommended,"—Science. 15 illustrations. 416pp. of text. 5⅝ x 8¼. 63702-6 Pa. $6.00

BURNHAM'S CELESTIAL HANDBOOK, Robert Burnham, Jr. Thorough, readable guide to the stars beyond our solar system. Exhaustive treatment, fully illustrated. Breakdown is alphabetical by constellation: Andromeda to Cetus in Vol. 1; Chamaeleon to Orion in Vol. 2; and Pavo to Vulpecula in Vol. 3. Hundreds of illustrations. Total of about 2000pp. 6⅛ x 9¼.
23567-X, 23568-8, 23673-0 Pa., Three-vol. set $26.85

THEORY OF WING SECTIONS: INCLUDING A SUMMARY OF AIRFOIL DATA, Ira H. Abbott and A. E. von Doenhoff. Concise compilation of subatomic aerodynamic characteristics of modern NASA wing sections, plus description of theory. 350pp. of tables. 693pp. 5⅜ x 8½.
60586-8 Pa. $6.50

DE RE METALLICA, Georgius Agricola. Translated by Herbert C. Hoover and Lou H. Hoover. The famous Hoover translation of greatest treatise on technological chemistry, engineering, geology, mining of early modern times (1556). All 289 original woodcuts. 638pp. 6¾ x 11.
60006-8 Clothbd. $17.50

THE ORIGIN OF CONTINENTS AND OCEANS, Alfred Wegener. One of the most influential, most controversial books in science, the classic statement for continental drift. Full 1966 translation of Wegener's final (1929) version. 64 illustrations. 246pp. 5⅜ x 8½. 61708-4 Pa. $3.00

THE PRINCIPLES OF PSYCHOLOGY, William James. Famous long course complete, unabridged. Stream of thought, time perception, memory, experimental methods; great work decades ahead of its time. Still valid, useful; read in many classes. 94 figures. Total of 1391pp. 5⅜ x 8½.
20381-6, 20382-4 Pa., Two-vol. set $13.00

MUSHROOMS, EDIBLE AND OTHERWISE, Miron E. Hard. Profusely illustrated, very useful guide to over 500 species of mushrooms growing in the Midwest and East. Nomenclature updated to 1976. 505 illustrations. 628pp. 6½ x 9¼. 23309-X Pa. $7.95

AN ILLUSTRATED FLORA OF THE NORTHERN UNITED STATES AND CANADA, Nathaniel L. Britton, Addison Brown. Encyclopedic work covers 4666 species, ferns on up. Everything. Full botanical information, illustration for each. This earlier edition is preferred by many to more recent revisions. 1913 edition. Over 4000 illustrations, total of 2087pp. 6⅛ x 9¼. 22642-5, 22643-3, 22644-1 Pa., Three-vol. set $24.00

MANUAL OF THE GRASSES OF THE UNITED STATES, A. S. Hitchcock, U.S. Dept. of Agriculture. The basic study of American grasses, both indigenous and escapes, cultivated and wild. Over 1400 species. Full descriptions, information. Over 1100 maps, illustrations. Total of 1051pp. 5⅜ x 8½. 22717-0, 22718-9 Pa., Two-vol. set $12.00

THE CACTACEAE,, Nathaniel L. Britton, John N. Rose. Exhaustive, definitive. Every cactus in the world. Full botanical descriptions. Thorough statement of nomenclatures, habitat, detailed finding keys. The one book needed by every cactus enthusiast. Over 1275 illustrations. Total of 1080pp. 8 x 10¼. 21191-6, 21192-4 Clothbd., Two-vol. set $35.00

AMERICAN MEDICINAL PLANTS, Charles F. Millspaugh. Full descriptions, 180 plants covered: history; physical description; methods of preparation with all chemical constituents extracted; all claimed curative or adverse effects. 180 full-page plates. Classification table. 804pp. 6½ x 9¼. 23034-1 Pa. $10.00

A MODERN HERBAL, Margaret Grieve. Much the fullest, most exact, most useful compilation of herbal material. Gigantic alphabetical encyclopedia, from aconite to zedoary, gives botanical information, medical properties, folklore, economic uses, and much else. Indispensable to serious reader. 161 illustrations. 888pp. 6½ x 9¼. (Available in U.S. only) 22798-7, 22799-5 Pa., Two-vol. set $11.00

THE HERBAL or GENERAL HISTORY OF PLANTS, John Gerard. The 1633 edition revised and enlarged by Thomas Johnson. Containing almost 2850 plant descriptions and 2705 superb illustrations, Gerard's Herbal is a monumental work, the book all modern English herbals are derived from, the one herbal every serious enthusiast should have in its entirety. Original editions are worth perhaps $750. 1678pp. 8½ x 12¼. 23147-X Clothbd. $50.00

MANUAL OF THE TREES OF NORTH AMERICA, Charles S. Sargent. The basic survey of every native tree and tree-like shrub, 717 species in all. Extremely full descriptions, information on habitat, growth, locales, economics, etc. Necessary to every serious tree lover. Over 100 finding keys. 783 illustrations. Total of 986pp. 5⅜ x 8½. 20277-1, 20278-X Pa., Two-vol. set $10.00

AMERICAN BIRD ENGRAVINGS, Alexander Wilson et al. All 76 plates. from Wilson's *American Ornithology* (1808-14), most important ornithological work before Audubon, plus 27 plates from the supplement (1825-33) by Charles Bonaparte. Over 250 birds portrayed. 8 plates also reproduced in full color. 111pp. 9⅜ x 12½.   23195-X Pa. $6.00

CRUICKSHANK'S PHOTOGRAPHS OF BIRDS OF AMERICA, Allan D. Cruickshank. Great ornithologist, photographer presents 177 closeups, groupings, panoramas, flightings, etc., of about 150 different birds. Expanded *Wings in the Wilderness*. Introduction by Helen G. Cruickshank. 191pp. 8¼ x 11.   23497-5 Pa. $6.00

AMERICAN WILDLIFE AND PLANTS, A. C. Martin, et al. Describes food habits of more than 1000 species of mammals, birds, fish. Special treatment of important food plants. Over 300 illustrations. 500pp. 5⅜ x 8½.
                                                         20793-5 Pa. $4.95

THE PEOPLE CALLED SHAKERS, Edward D. Andrews. Lifetime of research, definitive study of Shakers: origins, beliefs, practices, dances, social organization, furniture and crafts, impact on 19th-century USA, present heritage. Indispensable to student of American history, collector. 33 illustrations. 351pp. 5⅜ x 8½.   21081-2 Pa. $4.00

OLD NEW YORK IN EARLY PHOTOGRAPHS, Mary Black. New York City as it was in 1853-1901, through 196 wonderful photographs from N.-Y. Historical Society. Great Blizzard, Lincoln's funeral procession, great buildings. 228pp. 9 x 12.   22907-6 Pa. $7.95

MR. LINCOLN'S CAMERA MAN: MATHEW BRADY, Roy Meredith. Over 300 Brady photos reproduced directly from original negatives, photos. Jackson, Webster, Grant, Lee, Carnegie, Barnum; Lincoln; Battle Smoke, Death of Rebel Sniper, Atlanta Just After Capture. Lively commentary. 368pp. 8⅜ x 11¼.   23021-X Pa. $6.95

TRAVELS OF WILLIAM BARTRAM, William Bartram. From 1773-8, Bartram explored Northern Florida, Georgia, Carolinas, and reported on wild life, plants, Indians, early settlers. Basic account for period, entertaining reading. Edited by Mark Van Doren. 13 illustrations. 141pp. 5⅜ x 8½.   20013-2 Pa. $4.50

THE GENTLEMAN AND CABINET MAKER'S DIRECTOR, Thomas Chippendale. Full reprint, 1762 style book, most influential of all time; chairs, tables, sofas, mirrors, cabinets, etc. 200 plates, plus 24 photographs of surviving pieces. 249pp. 9⅞ x 12¾.   21601-2 Pa. $6.50

AMERICAN CARRIAGES, SLEIGHS, SULKIES AND CARTS, edited by Don H. Berkebile. 168 Victorian illustrations from catalogues, trade journals, fully captioned. Useful for artists. Author is Assoc. Curator, Div. of Transportation of Smithsonian Institution. 168pp. 8½ x 9½.
                                                         23328-6 Pa. $5.00

AMERICAN ANTIQUE FURNITURE, Edgar G. Miller, Jr. The basic coverage of all American furniture before 1840: chapters per item chronologically cover all types of furniture, with more than 2100 photos. Total of 1106pp. 7⅞ x 10¾. 21599-7, 21600-4 Pa., Two-vol. set $17.90

ILLUSTRATED GUIDE TO SHAKER FURNITURE, Robert Meader. Director, Shaker Museum, Old Chatham, presents up-to-date coverage of all furniture and appurtenances, with much on local styles not available elsewhere. 235 photos. 146pp. 9 x 12. 22819-3 Pa. $5.00

ORIENTAL RUGS, ANTIQUE AND MODERN, Walter A. Hawley. Persia, Turkey, Caucasus, Central Asia, China, other traditions. Best general survey of all aspects: styles and periods, manufacture, uses, symbols and their interpretation, and identification. 96 illustrations, 11 in color. 320pp. 6⅛ x 9¼. 22366-3 Pa. $6.00

CHINESE POTTERY AND PORCELAIN, R. L. Hobson. Detailed descriptions and analyses by former Keeper of the Department of Oriental Antiquities and Ethnography at the British Museum. Covers hundreds of pieces from primitive times to 1915. Still the standard text for most periods. 136 plates, 40 in full color. Total of 750pp. 5⅜ x 8½. 23253-0 Pa. $10.00

THE WARES OF THE MING DYNASTY, R. L. Hobson. Foremost scnolar examines and illustrates many varieties of Ming (1368-1644). Famous blue and white, polychrome, lesser-known styles and shapes. 117 illustrations, 9 full color, of outstanding pieces. Total of 263pp. 6⅛ x 9¼. (Available in U.S. only) 23652-8 Pa. $6.00

ACKERMANN'S COSTUME PLATES, Rudolph Ackermann. Selection of 96 plates from the *Repository of Arts*, best published source of costume for English fashion during the early 19th century. 12 plates also in color. Captions, glossary and introduction by editor Stel!a Blum. Total of 120pp. 8⅜ x 11¼. 23690-0 Pa. $4.50

Prices subject to change without notice.